WOMEN AND WASTELAND DEVELOPMENT IN INDIA

The ILO's World Employment Programme (WEP) aims to assist and encourage member States to adopt and implement active policies and projects designed to promote full, productive and freely chosen employment and to reduce poverty. Through its action-oriented research, technical advisory services, national projects and the work of its four regional employment teams in Africa, Asia and Latin America, the WEP pays special attention to the longer-term development problems of rural areas where the vast majority of poor and underemployed people still live, and to the rapidly growing urban informal sector.

At the same time, in response to the economic crises and the growth in open unemployment of the 1980s, the WEP has entered into an ongoing dialogue with the social partners and other international agencies on the social dimensions of adjustment, and is devoting a major part of its policy analysis and advice to achieving greater equity in structural adjustment programmes. Employment and poverty monitoring, direct employment creation and income generation for vulnerable groups, linkages between macro-economic and micro-economic interventions, technological change and labour market problems and policies are among the areas covered.

Through these overall activities, the ILO has been able to help national decision-makers to reshape their policies and plans with the aim of eradicating mass poverty and promoting productive employment.

This publication is the outcome of a WEP project.

WOMEN AND WASTELAND DEVELOPMENT IN INDIA

EDITED BY

Andrea M. Singh
Neera Burra

A study prepared for the International Labour Office within the framework of the World Employment Programme.

Sage Publications
New Delhi/Newbury Park/London

First published in 1993 by

Sage Publications India Pvt Ltd
32 M–Block Market, Greater Kailash–I
New Delhi 110 048

Sage Publications Inc
2455 Teller Road
Newbury Park, California 91320

Sage Publications Ltd
6 Bonhill Street
London EC2A 4PU

Published by Tejeshwar Singh for Sage Publications India Pvt Ltd, phototypeset by Vijaya Enterprises, Madras and printed at Chaman Enterprises, Delhi.

Library of Congress Cataloging–in–Publication Data

Women and wasteland development in India / edited by Andrea M. Singh, Neera Burra.

p. cm.

Papers presented at the National Technical Workshop on Women and Wasteland Development, organized by the International Labour Organisation, held in New Delhi, Jan. 1991.

Includes bibliographical references (p.) and index.

1. Women in rural development—India—Congresses. 2. Women in agriculture—India—Congresses. 3. Women in conservation of natural resources—India—Congresses. 4. Women and land use planning—India—Congresses. 5. Reclamation of land—India—Congresses. 6. Waste lands—India—Congresses. I. Singh, Andrea Menefee. 1943– II. Burra, Neera, 1951– III. National Technical Workshop on Women and Wasteland Development (1991: New Delhi, India)

HQ1240.5. I4W65 1993 305.42.0954—dc20 93–9020

ISBN 0–8039–9123–1 (US)
81–7036–349–7 (India)

Contents

Foreword

The papers presented in this volume were prepared for the 'National Technical Workshop on Women and Wasteland Development' that was organised by the International Labour Organisation (ILO) in New Delhi in January 1991, with the financial support of DANIDA (Danish International Development Agency) and the Royal Netherlands Government. The Workshop brought together approximately seventy grass-roots-level activists, policy-makers, researchers, government officials and representatives of international and donor agencies to discuss and disseminate different approaches to involving women in wasteland development.

The vicious circle of poverty, environmental degradation and rural livelihoods has received increasing attention in the last few years. There has also been a better understanding and recognition of the central role women play in this process. Specific references in the Agenda 21 of the Earth Summit held in Rio are witness to this.

The experiences presented in this volume were mostly initiated in the early eighties and go beyond discourse to assess concrete actions taken to enable women's control over management of environmental resources. The rich debate on important—and at times controversial—issues, and the variety of approaches and institutional experiences presented in the papers, yield important lessons for planners, policy workers and grass-roots organisations.

The Workshop was organised by the ILO as a policy debate forum within the framework of its programme on 'Women and Wasteland Development in India' initiated in the mid-eighties. This programme was an outcome of the dynamic interrelationship between policy research and field-level technical cooperation of the ILO's 'Programme on Rural Women'. Established in the mid-seventies, the Programme had focused its activities on providing a better understanding of patterns of employment and poverty among women workers, the impact of policies and programmes on women's position in the labour process, their responses and initiatives. These have formed the basis for advice on policies, preparation of comprehensive action plans and implementation of pilot schemes.

Among the main areas of focus of the programme in the eighties were those of women, poverty and livelihood and environmental deterioration. Global research was supported in the early eighties on rural women's energy needs and emerging alternatives. A comprehensive study was carried out on the historical dimensions of women's rights to land and land-based resources in South Asia. A review of legislation examined the impact of personal laws, colonial forest policies and traditional rights of tribal peoples. At the same time, on the ground, technical assistance was being provided to an organisation of rural women workers, mainly tribals and agricultural wage labourers, in the Bankura district of West Bengal. The absence of sufficient resources for local employment forced many of these women to engage in migrant agricultural labour two or three times a year. Initially, various off-farm activities and small animal husbandry activities were promoted in recognition of women's lesser access to or control over land-based resources, but the impact of these activities was insufficient to cope with the problem. The involvement of women in wasteland development was an abundant, albeit degraded, resource in this area and came from the women themselves and their supporters. Awkwardly at first, and largely through a process of trial and error, thriving plantations eventually emerged.

This has now created a movement which has received recognition as a strategy to regenerate the environment, significantly increase employment opportunities through an expanded resource base, and to help make labour migration a matter of choice. This experience as well as other schemes reviewed in this volume illustrate how pilot action programmes could lead to policy debate and provide effective approaches for initiating policy reform.

Azita Berar Awad
Employment and Development Department
International Labour Organisation

Acknowledgements

The book is the outcome of a National Technical Workshop on 'Women and Wasteland Development' that was organised by the International Labour Organisation (ILO) in New Delhi, 9–11 January 1991. It contains a selection of papers that were commissioned for the Workshop. The introduction brings out the major issues that were highlighted in the papers. The conclusions of the Workshop are reproduced in the last chapter.

We are indebted to all the paper writers and to all the participants of the Workshop who contributed to making the Workshop a success. We hope that this publication will throw some light on a complex subject and that it will stimulate both research as well as direct action plans for women's participation in natural resource management.

We wish to acknowledge the interest shown in this work by the International Labour Organisation and the support it extended towards its publication.

We would also like to express our gratitude to Ms. Azita Berar Awad of the ILO for her overall support regarding the issue of women and wasteland development in general and the present endeavour in particular. Of special importance has been the financial support of DANIDA (Danish International Development Agency) and the Royal Netherlands Government for the conduct of the Workshop.

We are grateful to Ms. Kokila Singh, Ms. Harmeet Sarin and Ms. Prema Viswanathan for the time and trouble they have taken in typing the papers. Ms. Viswanathan, in particular, has with her usual good cheer spent many hours in preparing the final manuscript.

November 1992

Andrea Menefee Singh
Neera Burra

Programme on Rural Women
Policies and Programmes for Development Branch
Employment and Development Department
International Labour Organisation

Introduction

ANDREA M. SINGH
NEERA BURRA

What is wasteland? Current literature reveals several definitions. One definition is: 'lands which are either lying entirely barren or are producing much below their economic potential' (PRIA, 1984: 4). A more comprehensive definition of wasteland is offered by Robert Chambers, N.C.Saxena and Tushaar Shah (Chambers et al., 1989). According to their study:

> Wastelands mean degraded land which can be brought under vegetative cover, with reasonable effort, and which is currently lying underutilised and land which is deteriorating for lack of appropriate water and soil management or on account of natural causes. Wasteland can result from inherent/imposed disabilities such as by location, environment, chemical and physical properties of the soil or financial or management constraints. (ibid.: 42)

In India there are three types of public wastelands as described by Chambers et al. (1989). These are forest wastelands, wastelands which are owned by the government but used by the community and grazing lands which are controlled by village bodies. Thus there are wastelands under the Forest Department, wastelands under the control of the Revenue Department and common grazing lands which are supposed to belong to the village but which have been largely encroached upon.

There are vast tracts of wastelands belonging to different government departments. There are an estimated 55 million hectares of uncultivated non-forest lands of which 35 million hectares consist of fallow and owned/allotted/encroached land which is now under private control. The rest of the 20 million hectares is still under the control of the Revenue Department (Chambers et al., 1989: 44). Chandra (1990: 33) estimates that of India's total land mass of 328 million hectares, nearly 61 million hectares are wastelands.

The regeneration of wasteland in India has, for some years, attracted the attention of governmental policy and programmes, NGOs and international agencies. However, the question of the participation of women in the development of wasteland, and its impact upon their lives has acquired significance more recently. In the last few years, the initiatives of the ILO in Gujarat and West Bengal and the efforts of NGOs in different parts of the country have yielded many insights into the complexities of developmental programmes that seek to involve women in wasteland development. By 1991, the ILO felt that the time for stocktaking had arrived because a number of persistent issues had arisen out of field-level experience.

What should be the type of land selected for women's involvement? What are the legal difficulties in securing the rights of women over private, forest, revenue or *panchayat* lands? To what extent should such programmes exclude men and how is their potential hostility to be countered? If women are largely involved as wage workers in wasteland programmes, what type of organisation would best protect their rights—a cooperative, an informal women's group, a *mahila mandal* or something else? And what are the legal implications in terms of securing the lease or ownership of land? These and similar types of questions began to demand some answers. The ILO organised a National Technical Workshop on 'Women and Wasteland Development' in January 1991. One purpose of the Workshop was to bring together people who were concerned with these issues from different perspectives—the field, the government, academia and funding agencies—in the expectation that some light would be shed on these matters.

Martha Chen's background paper traces the origins and development of government policies regarding wasteland development and observes that the earlier social forestry programmes of the Government of India tended to benefit larger farm households rather than smaller farm (or landless) households, and promoted timber production rather than fuelwood and fodder. It looks at the many facets of the relationship between women and wasteland and suggests a methodology to enhance the participation of women. Chen's paper attempts to integrate women into a wider analytical framework for promoting wasteland development and argues that women's participation—including women's knowledge, responsibilities, and rights (both legal and political)—should be seen as integral to the process, indeed to the success, of wasteland development. She looks at about 20 projects in eight different states covering several different agro-ecological regions, and she draws upon that rich comparative experience in her paper.

When one looks at the experience of a decade or so, it is apparent that what once seemed a simple task has turned out to be far more complex than imagined. Chen notes that, while the earlier response was to simply locate some land and get some people to plant trees on it, the language of discourse has changed significantly now. And so too have the concepts. One noticeable difference between governmental and NGO efforts has been that while the former have tended to concentrate on targeting individual households, the latter have favoured group activity. A great debate also, now rages over the question of the 'privatisation of the commons' in situations where a select group is given exclusive rights over land which formerly was accessible to the community at large. It is worth mentioning here that some experts and activists in the field have endorsed the notion that traditional user groups should be restored their rights over their land.

Chen goes on to point out that if the involvement of local people was sought earlier, it was largely as members of the labour force, but now, most wasteland development projects seem to have clearly crystallised their economic, distributive and environmental objectives. How women are perceived and treated has also undergone significant change. At one time, they were seen as the primary victims of environmental degradation; yet, in practice, they worked only as wage labour on nurseries and plantations. As an understanding of their role in the household to mediate the adverse consequences of denudation emerged, their place in wasteland projects underwent a transformation. Many of the contributors to this volume are of the opinion that, given both their knowledge of the natural resource base and their orientations, women's involvement in designing wasteland projects and in choosing species for plantation is crucial for the welfare of the household. In the absence of title to land or access to other inputs, it is clearly an uphill task to bring women to the centre of the stage in a male-dominated society. This raises the question of how to provide women with the wherewithal to gain legal rights over land and give content to those rights through a suitable institutional framework. Chen urges that only a strategy that answers this question will successfully meet the practical, strategic and empowerment requirements of women.

If one looks at the case studies of the People's Education and Development Organisation (PEDO) in Rajasthan by Madhu Sarin and Chandrika Sharma and of Social Action with Rural and Tribal Inhabitants of India (SARTHI) in Gujarat by Madhu Sarin and Renu Khanna, it becomes apparent that by targeting women exclusively, suspicion and resentment among men in the community are likely to grow. The studies indicate that holding joint meetings initially with both men and women can openly address such fears and help to create a non-threatening atmosphere in which women's groups can function freely and without inhibition. Both case studies point to the need for holding regular group meetings to raise awareness and to increase the confidence of women in a conservative social structure. Individual women find it difficult to break out of their shells without the strength of the collectivity behind them. Both case studies suggest that though women are the prime sufferers on account of environmental degradation, it is not easy for them to become organised and mobilised. Women may be inspired to take up wasteland development for economic reasons and to reduce the time and drudgery involved in fodder and fuel collection, yet initially they are likely to face strong opposition in their own families and in society at large. An outside catalyst, such as an NGO, can play an important role in raising awareness of the issues and galvanising the energies of women. It was brought out in the case study of SARTHI that one of the biggest problems is the ambiguous legal status of the land being developed. There is considerable uncertainty regarding the validity of their title and great concern as to whether they would be able to legally secure their long-term rights to the land being developed.

The paper on the role and participation of women in Forest Protection Committees (FPCs) in West Bengal by Philip Viegas and Geeta Menon describes the historically antagonistic relationship between the Forest Department and the people who are dependent upon the forests. The authors note how the commercialisation of forest resources and the process of industrialisation have placed great strain upon forest-dwelling communities as the denudation of forests proceeds apace. The novel experiment of FPCs is an attempt to reverse this trend and looks at people as partners in the management of forests rather than as destroyers of forests. FPCs have become very popular in West Bengal because the people can freely collect non-timber forest produce (NTFP) and are entitled to 25 per cent of the value of timber at the time of clear felling.

The paper studies 10 FPCs operating under different conditions and analyses the factors that impede or promote women's participation. The authors found that in the absence of formal representation for women, their participation was low or non-existent. But even where there was formal representation, there were cases where 'the culture of silence' inhibited women in taking part in the decision-making process. However, and this is noteworthy, where voluntary agencies took the lead by organising women, their participation in FPCs was markedly more pronounced. The authors also observed that where women played an effective role in FPCs their status rose within their community and feelings of solidarity amongst women developed. If earlier a woman bore her physical and psychological burdens alone, today she can at least depend to some extent on her female friends. The authors conclude on a cautionary note that the future of FPCs, though potentially bright, will be limited by both the Indian Forest Act and any forest policy which promote industries rather than people's rights. The current policy still looks upon the people as being responsible for the present condition of forests.

Vasudha Dhagamwar and Enakshi Ganguly Thukral were asked by the ILO to study the issues concerning women's involvement with wasteland development in the state of Gujarat. While they review the projects of a few other agencies in their paper, they concentrate upon the ILO/SEWA wasteland project and examine the relevant legislation, rules and government orders governing the grant of rights over wasteland concerned. One conclusion that emerges from the paper is that it is extremely difficult to satisfy the stringent conditions of both State and Central legislation. An important point made by the authors is that there are mutually contradictory requirements laid down by the Forest Department and the Registrar of Cooperative Societies. While the former insists that a cooperative be registered before land can be granted, the latter insists that land be granted before the cooperative is registered! On the basis of their study of women's groups involved in wasteland development in different villages, Dhagamwar and Thukral argue that many of the problems faced by these groups are not just legal in nature, but stem from competition for scarce land resources, entrenched caste rivalries and the political machinations of village-level elected bodies. The authors draw attention to the fact that the lease period, that is, 15 years in Gujarat, is too short for the fruits of afforestation to be enjoyed by the women who worked the land. They suggest that either the leases should be for 50 years or renewed for 15 years at a time. Moreover, excessive discretion is placed in the hands of the relevant authorities who can permit cancellation of leases for almost any reason. They underline the need for renewal of leases to be the norm as also for laying down in greater detail the conditions under which leases would merit cancellation. Their study stresses, as do so many of the other papers in this volume, the importance of NGOs who liaise with governmental authorities, pick up the legal gauntlet where necessary and generally handle paperwork on behalf of the village women. What emerges from the paper is the need for poor, rural and largely illiterate women to organise and draw strength from their collectivity.

Neera Burra's paper reviews the experiences of NGOs with women's involvement in wasteland development. It is based on visits to a large number of NGOs in the country which are actively engaged in wasteland development work. It attempts to place women's participation in wasteland development in the larger sociological context of village society by describing and analysing the different approaches adopted by NGOs. Caste and class, male out-migration, relative homogeneity of villages and topography have been identified as some of the factors which have a strong bearing on women's successful participation in wasteland development. There are also marked regional variations and these have been documented in her paper. The author delineates the circumstances and contexts that promote or discourage women's participation in wasteland development.

Most of the papers in this volume argue that women's needs and priorities are different from those of men. It has been claimed by several contributors that while men usually want to grow commercial species, women prefer the broad-leaved species which meet their subsistence needs for fuel and fodder. Burra, however, finds that although there is a case for organising wasteland development projects with women as the target group, many of the projects she studied did not see a difference between the needs of men and women. She argues that the implementation of development programmes in general certainly needs a sharper focus upon women in order to enhance their income-earning opportunities and status—both of which are considerably lower in most of India when compared to those of men. The underlying conflict, she suggests, is really between the needs of the poor on the one hand, and those of commerce and industry on the other.

N.C. Saxena's paper on the policy issues relating to women and wasteland development differs from the other papers in this volume insofar as it adopts a macro-perspective towards the problems involved. He argues that individual efforts by NGOs, praiseworthy as they may be, can but tinker with the problem and only radical policy changes at governmental level can have a far-reaching impact on the poor and, amongst them, more particularly, women. He observes that social forestry projects in India have hardly considered the role of women at either the project design or implementation stage. Women have remained 'invisible' to planners and government officials, and have been 'designed out' of projects by default.

The paper suggests that social forestry projects should be extended to reserved and protected forest lands by changing the nature of species from timber-oriented trees to usufruct-based trees. These should be supplemented with shrubs and bushes to yield fuelwood and fodder in the shortest possible time. This would strengthen women's access to forests and, therefore, benefits would be directly appropriated by them. The strategy should be to opt for species which have high proportions of branches and twigs relative to stem wood.

He points out that such a strategy would require a complete reversal of the recommendations of the National Commission on Agriculture, 1976, which favoured the plantation of commercial species on forest land, and trees for consumption and subsistence on private land. What is suggested in the paper is that subsistence needs and consumption demand should be met from forest and common lands, and commercial and market demand should by and large be met from private land. Saxena's suggestions, if they were to be implemented, would require amendment of both the

New Forest Policy and the Forest Conservation Act. He calls for genuine people's participation by permitting villagers to share in the management of forests and to grow those species that would meet the needs of the poor.

One recurrent theme of the volume is the need for women to organise and gain strength from the whole: one cannot expect women in a male-dominated society to individually seek or gain their liberation, particularly given the contexts in which they are socialised. Another conclusion that features in many of the contributors' writings is that a catalyst in the form of an external or intermediary NGO is perhaps necessary to generate the confidence needed for women to join and build an organisation of their own. Ordinary poor, illiterate, rural women would not appear to have the resources or the experience to deal directly with *Panchayats*, bureaucracies, governments and law courts, at least not initially.

The uncertain legal status of the land on which women are working in many projects may turn out to be a major obstacle to long-term sustainable development. It is not at all clear whether some of the women described in this volume will be allowed to retain control over their land indefinitely or, where there are leases, when their leases expire. Legal hurdles in giving women access to land are multiple because the grant of different types of land is governed by different sets of rules. Moreover, several obstacles in registering cooperatives or other organisations that enjoy legal status need to be removed. The electoral politics of caste and class at the village-level and the overwhelming dominance of the male in society are factors that do not augur well for projects or programmes that seek to link women with the development of wasteland. But an understanding of the social context, is surely the first step in overcoming the hurdles it embodies. There are stories in this volume of both failure and success, and each account offers its own lessons for the future.

The last chapter in this volume embodies the conclusions of the workshop. These were discussed and debated in full house sessions and represent the consensus arrived at by the participants on major policy and programmatic concerns. Eight peasant women from eight different villages and representing four states of India took part in the workshop as full participants. Arrangements were made for all the presentations and discussions to be translated into the four Indian languages these women spoke. As a result, their participation could be wholehearted and they raised a number of questions and offered their comments through the course of the Workshop. The presence and effective contribution of these village women were the highlight of the proceedings as they spoke from the reality of experience and engaged in uninhibited dialogue with all those present.

This book contains a selection of the papers presented at the Workshop. We would like to thank all the participants at the Workshop and in particular all those who wrote papers. We have sought to present as faithfully as possible the different viewpoints and opinions expressed at the Workshop.

As organisers of the Workshop and editors of this volume, we hope that these papers will contribute in at least some measure to illuminating the issues that concern women and wasteland development.

REFERENCES

Chandra, Ramesh (1990) 'Role of Panchayats as Extension Agents of Wasteland Development', in Ramesh Chandra (ed.), *Role of panchayat in social forestry and wasteland development*. New Delhi: National Mission on Wasteland Development, Ministry of Environment and Forestry.

Chambers, Robert, N.C. Saxena and Tushaar Shah (1989) *To the hands of the poor: Water and trees*. New Delhi: Oxford and IBH.

PRIA (1984) *Women and wasteland development*. New Delhi: Society for Participatory Research in Asia.

1 *Women and Wasteland Development in India: An Issue Paper*

MARTHA ALTER CHEN

The Context

> The fundamental cause of poverty in India arises out of the scarcity of biomass resources to meet daily basic needs like food, fuel, fodder, manure, building materials and artisanal raw materials, almost all of which are biomass-based. (Agarwal and Narain, 1989: viii)

> In the traditional division of labour between men and women in a predominantly subsistence economy, 'women's work' of collecting cooking fuel, fetching water and cattle care binds them more intimately with the natural resource base than men. It is they who have to pay the price for environmental degradation by having to walk longer distances in their daily search for fuel, fodder and water. (Sarin and Khanna, 1990: 6)

Land use in India

The land resources of India are under constant and increasing pressure. The combined pressures of population growth, agricultural expansion, and industrial growth have led to widespread environmental degradation and to significant changes in land use patterns. Between 30 and 50 per cent of India's land area is wasted (Agarwal and Narain, 1989:1). As land-based resources become scarce or more degraded, the balance between what people need and what they can obtain shifts. For rural families, the struggle to survive becomes harder. The effect on women is particularly severe because women are the primary collectors and users of natural resources in most parts of rural India.

Since 1900, the population of India has more than tripled. Agricultural development has been equally dramatic. In the first half of this century, the arable area was rapidly expanded (by as much as 45 per cent in many northern districts) while forests, woods, grasslands, and scrublands were reduced (by 30 to 40 per cent in the same areas).[1] And the irrigated acreage expanded ninefold between 1890 and 1970 (Repetto, 1987: 18–19). Since 1950, there has been little expansion of gross cropped area but production has continued to increase due to widespread adaptation of high yielding technologies. During this century, the industrial sector has also grown dramatically, placing additional pressures on land-based resources.

In the past, the pressure on India's land base was discussed largely in terms of increases in crop yields on private lands relative to increases in population. Today, this pressure is discussed widely in terms of the productivity of biomass on common lands relative to the subsistence needs of people and the material needs of industry. Indian environmentalists have introduced this change in thinking. Their studies have shown that, in India, the majority of people subsist on a biomass-based economy and almost half the industrial output comes from biomass-based industry (Agarwal and Narain, 1989:1). The need to increase biomass production has prompted increased attention to trees and grasses (in addition to standard crops) and to India's forests and grasslands (in addition to its croplands).

Behind this change in thinking is the growing realisation that poverty and environmental degradation are two sides of the same coin (Chowdhry, 1988: 21). The close connection between poverty and land degradation has several dimensions. Foresters and planners have generally viewed those who live near forests, particularly

[1] In Dungarpur district in southern Rajasthan, for example, there has been a significant expansion in cultivated area and decline in forest area (Sarin and Sharma, 1990):

	1943–44	*1985–86*
Population	286,000	682,000
Cultivated Area	41,300 ha	130,097 ha
Forest Area	332,348 ha	61,815 ha

the poor, as a major cause of land degradation and deforestation. There is no doubt that forests and grazing lands are major resources for local populations. But there is also no doubt that local populations are only one of several competitors, including government agencies and industrial concerns, for these resources. And that the linkages between poverty, population growth, economic growth, and land degradation are complex.

Although the estimates differ, about 260 million hectares of India's land are available for biomass production.[2] About 64 per cent of this amount is private land—agricultural fields, groves and plantations, or fallow. About 32 per cent is under the control of state governments—roughly 25 per cent under forest departments and six per cent under revenue authorities. However, a large percentage of the revenue land has been encroached upon and, thereby, effectively privatised. About five per cent is under the control of village *panchayats* and is used predominantly as grazing lands.

To date, biomass production has increased on only a very small portion of the total: that is, on irrigated croplands. On the rest, productivity is still very low or even declining. Indeed, between one-third and one-half of this land can, in ecological terms, be called wasteland (Agarwal and Narain, 1989: 1). The most degraded and underproductive lands are those under the control of *panchayats*, revenue authorities, and the forest departments. The production of biomass on these various categories of land has been summarised as follows:

—the productivity of forest lands has been steadily decreasing—due to ongoing erosion of the forest cover;
—the productivity of grasslands is a mere fraction of their biological potential—in most cases less than 20 per cent;
—the productivity of croplands has increased mainly in the irrigated areas (ibid.: 1).

Although land is the primary physical unit of wasteland development, it is not the only unit. Another important physical resource is water, in the form of ponds, tanks, *nalas* (gullies), or watersheds. In recent years, India's environmentalists have begun

[2] Of India's roughly 300 million hectares of land area, some 40 million hectares 'is rocky, under glaciers or under settlements and is, therefore, not available for growing any form of biomass' (Agarwal and Narain, 1989: 27).

to focus on developing watersheds. This involves comprehensive integrated planning and water or soil conservation, and not simply land development. Environmentalists have also introduced the ecosystem as a unit of analysis: that is, the system which connects all biological life with the physical resources upon which it depends. India is seen as a country with extremely diverse ecosystems, each of which can produce and sustain a specific kind of biomass. Wasteland development must, according to this analysis, be ecosystem-specific.[3]

Under this expanded definition of wasteland development, trees are only one of several sources of biomass. Further, the value of trees as timber (dead wood) may be far less significant than the value of trees as a source of non-timber forest products (NTFP).[4] Other sources of biomass to be produced or protected include grasses, bushes, shrubs and agricultural by-products.

Each wasteland development project must study and take into account local social and ecological dynamics, existing ownership and usufruct systems, and the exact nature of the problem to be addressed. Each physical unit may be affected by one or more of the following problems:

—decrease in quantity (due to deforestation, privatisation or other factors);
—decrease in quality (due to soil and water loss, siltation, overgrazing, or other factors); and/or
—change in access or use patterns (due to increased competition, abuse of customary rights, abuse of concessionary agreements, or other factors).

[3] Not only regions of India but also individual villages have distinct ecosystems. Each village can be seen as a complex system in which several sub-systems interact, including land, water, livestock and energy (Agarwal and Narain, 1989: 8ff). Women play important roles in linking and managing the energy flows from one sub-system to another.

[4] In the past, non-timber forest products (NTFP) were commonly referred to as minor forest products (MFP). But in recent years, the value of these products has been measured and seen to be significant, not minor. Hence the new term for them. Another term used in the literature is non-wood forest products (NWFP), but NTFP is used in this paper.

Fuelwood and Fodder Crises

The competing pressures on India's land resources have led to two interrelated crises in India: the fuelwood crisis and the fodder crisis. In India as a whole, woodfuel (fuelwood and charcoal) is estimated to be about 33 per cent of the total energy produced, both commercial and woodfuel (Agarwal, 1986a: 8). One rural study reports that fuelwood accounts for over 80 per cent of the total energy produced and that 96 per cent of fuelwood is consumed in the home—82 per cent for cooking and 14 per cent for heating water. Other rural studies indicate that fuelwood constitutes almost 69 per cent of total energy: of which less than 13 per cent is purchased, 23 per cent is home grown, and almost 65 per cent is freely collected (Kaur, 1988: 21).[5] The National Commission on Agriculture estimates that consumption of fuelwood in India is about four times the production (Singh, 1990: 2).

There are, of course, other sources of rural domestic energy, including: twigs or sticks, dung and agricultural residues (corn cobs, sugarcane husks, rice straw). Shortages in these resources have also contributed to the rural energy crisis. And shortages of these other biomass fuels can also be attributed to competition between local populations, agricultural development, and industrial growth. For instance, one study argues that the shortage of animal dung is due to three basic reasons: agricultural expansion on grazing lands; increasing use of animal dung as manure; and increasing use of animal dung and crop residues in gasifiers (Maiti and Dasgupta, 1986: 7).

In India, livestock is fed overwhelmingly on crop residues and natural forage, rather than on fodder crops. And livestock has increased in numbers proportionately with the expansion of cropped area (Repetto, 1987). This increase in livestock population has put increased pressure on the remaining non-arable land. In many areas, both in the plains and the hills, where livestock (and dairying) are an important component of household economies, fodder is scarce as compared to fuelwood.

These crises have affected the rural poor disproportionately.

[5] Fuelwood or firewood consists of split logs and large tree branches (Kaur, 1988: 24). Woodfuel consists of fuelwood and charcoal (Agarwal, 1986a: 8).

Because they have been largely excluded from private property, the poor depend heavily on the free collection of fuelwood and fodder from common property resources (CPR) land. A study in seven states found that poor households graze their animals on the commons from 70 to 90 per cent of the time, compared to 20 to 40 per cent for better off rural households. Poor households obtain 60 to 90 per cent of their fuel from the common lands, compared to 10 to 20 per cent for other households. The rural poor also rely much more heavily on common land for gathering food, raw materials for artisan production, and other subsistence needs (Jodha, 1986).[6] And among the rural poor, these crises have affected rural women disproportionately because women are the primary collectors, users, and managers of fuelwood and fodder.

Social Forestry and Wasteland Development

As early as 1973, Indian foresters and planners introduced the concept of 'social forestry' to meet the 'social demands on forests', defined as follows: fuelwood supply to the rural areas and replacement of cow dung; small timber supply: supply of grasses and fodder and provision of grazing; protection of agricultural fields against wind; and recreational needs (GOI, 1973). Over the next several years, the Government of India supported a variety of schemes to promote fuelwood, fodder, and small timber production. But the early social forestry programme had several weaknesses. Most notably, the programme exceeded its targets in promoting individual households to grow trees on their private land but fell short of its targets in promoting village communities to grow trees on common lands (Cernea, 1989; Molnar and Schreiber, 1989). As a result, the schemes tended to benefit larger farm households rather than smaller farm or landless households and to produce more timber than fuelwood and fodder.

Partly in response to weaknesses in the early social forestry programme and largely in response to the need to regenerate degraded common lands, a wasteland development movement was

[6] This study also estimated that, in 1951, the per hectare dependence on CPR land was 4.9 persons. By 1981, it had increased to 13.7 persons.

initiated in the early 1980s. Unlike the early social forestry programmes, the wasteland development movement focused more on common than on private lands, promoted more fuelwood and fodder than timber plantations, and worked more closely with local communities and non-government organisations (NGOs) than with individual households. In 1985, in response to this movement, the Government of India constituted a National Wasteland Development Board (NWDB), as a semi-autonomous body. The task of formulating a social forestry policy was shifted from the Ministry of Environment and Forests to the NWDB, which was charged with the mandate of bringing five million hectares of land each year under fuelwood and fodder plantation and developing a people's movement for afforestation. Recently, the NWDB was reconstituted as a Technology Mission of the Government of India.

During the 1980s, there was an active debate on how best to promote social forestry or wasteland development. Initially, foresters and planners advocated a community approach to afforestation but defined 'community' in vague and ambiguous terms.[7] Evaluations of these early efforts reported widespread failure of the community approach, due to heterogeneity of the communities, control by local elite, breakdown of indigenous systems of community management, and other factors (Molnar, 1986). Some foresters and planners, therefore, began to favour working with individual households, rather than with 'communities'. Many environmentalists and NGOs reacted against the family or household approach, characterising it as 'privatisation', and promoted group-based approaches which focused on clearly identified user groups.

Several of these recent efforts at group management of common lands have been successful, including several reviewed for this study. Evaluations of these efforts indicate a common set of characteristics, which were largely missing from the original models for establishing 'community' woodlots, notably: management that rests with a local user group, building on both customary

[7] The term environmentalist is used here to refer to researchers and activists who work on environmental issues. Those in government who work on these issues are referred to as foresters, if they work in Forest Departments, or as planners, if they work elsewhere in government. The term women's movement is used here to refer to researchers and activists who work on women's issues.

and formal institutional arrangements; publicly acknowledged rights and duties of the users; a distribution system that reaches a diverse population; and a protection system that has clear and easily enforced rules of compliance (Molnar, 1986).

The literature on social forestry, wasteland development, and common property resource management provides useful insights on women's traditional roles as collectors of fuel, fodder, and other non-timber forest produce but does not fully integrate women into the wider analytical framework for promoting wasteland development. This paper attempts to do so and argues that women's participation—including women's knowledge, responsibilities, and rights (both legal and political)—should be seen as integral to the process, indeed the success, of wasteland development.

Issue Paper

This paper summarises women's roles in relationship to natural resources in India and presents a rationale and methodology for promoting women's participation in wasteland development projects. In doing so, the paper attempts to synthesise and generalise from complex and diverse project experience. Twenty projects in eight states were reviewed, reflecting the differences between four distinct agro-ecological regions (arid or semi-arid plains, hill, degraded forest fringe, standing forest fringe).[8]

The paper is divided into four main sections. Section I reviews the various roles women (and men) play in regard to natural resources.[9] Section II summarises five approaches to wasteland

[8] In early 1990, Neera Burra of the ILO and I visited ten projects in four States (Gujarat, Rajasthan, West Bengal, and Orissa). Later, Neera Burra visited an additional seven projects in another four States (Uttar Pradesh, Tamil Nadu, Karnataka, and Andhra Pradesh). In early 1990, the ILO commissioned four case studies of projects in four States (Gujarat, West Bengal, Rajasthan, and Uttar Pradesh) plus two legal studies. These projects, listed in Appendix 1.1, are used, together with other documented projects, as illustrative materials throughout the paper.

[9] Cf. Kaur, 1988 for a detailed review of the literature on women in forestry in India, and Molnar and Schreiber, 1989 for various operational issues (planning issues, design features, and information needs) relating to women and forestry projects.

development (private management of private lands; and private, group, government, or joint management of common lands) and analyses the ways in which local people have been involved. Section III explores why women (in particular) should be involved, how they have been involved, and the impact of doing so. The final Section IV presents a framework for analysing women's participation in wasteland development and discusses three central strategic needs of women: property rights; political representation; and gender-based land use planning.

In this paper, the terms wasteland, common property resources, wasteland development, and biomass are used in the broad and inclusive sense, as follows:

— Wasteland: ecologically degraded land, private or common.

—Common property resources: all water and land resources to which local people at one time had common access, including private agricultural land under seasonal fallow, village *panchayat* land, revenue lands, and protected forest lands.[10]

—Wasteland development: efforts to generate or regenerate biomass on ecologically degraded land (including afforestation, regeneration through protection, watershed development).

—Biomass: products from plants and animals, not just trees, including food, fuel, fodder, manure, building materials, artisanal raw materials, plant fibres, medicinal plants and herbs, fruits and nuts, seeds, oils and resins, and other non-timber forest products.

At times, natural resources or, simply resources are used as a general term for all land-based resources, both the different types of land themselves and the range of biomass products. Finally, unless otherwise indicated, the term women refers to rural women who collect biomass resources. As an important variable of analysis, the paper will distinguish, as necessary, between groups of rural women by four types of habitat or agro-ecological zones: those who live near degraded forests, near standing forests, in the hills, and in arid or semi-arid plains And, when necessary, the paper will categorise women by caste, occupation, or other characteristics.

[10] Common property resources (CPR) are defined here in an inclusive usufruct sense, not in a political or legal sense. For a similar application of the term , cf. Arnold and Stewart, 1989.

Section I: Women and Natural Resources

There are several sets of studies on women and natural resources: those that study the connection between women and the fuelwood crises (e.g., Agarwal, 1986a); those that study women's relationship to forests (e.g., Shiva, 1988; Fernandes and Menon 1987); and a few that study women's relationships to common property resources other than forests (e.g., Chen, 1991). The majority of early studies focused on the consequences of deforestation on women's time allocation, documenting the increasing amounts of time women spend collecting fuel and fodder as forests are degraded. The outcome of this preoccupation was the well-publicised but stylised picture of women as the primary gatherers of fuelwood and fodder for household consumption.

Recent studies have described a more complete picture, including: first, the range of products that women collect from grasslands and fallows as well as from forests; and, second, the range of women's roles and responsibilities in regard to these products. This more complete picture depicts women as users and managers of a range of products in both the domestic and market economies. The other biomass products collected by women include: fruit, herbs and plants for local medicines, flowers for oil, liquor, cattle feed; leaves, seeds, and pods of various trees; various types of fibres and floss, tans and dyes (Agarwal, 1986a: 23–24). The biomass-related roles that women undertake in addition to gathering include processing and storing, utilisation and management, and marketing.

Women, Resources, and the Domestic Economy

The linkages between women, natural resources, and the domestic economies of poor rural households in India are fairly clear. Recent studies have confirmed three common patterns. First, natural resources are central to the livelihoods of poor rural households in India. Many studies, including the seven-state study noted earlier, have documented that poor households are more dependent on free biomass goods and common property resources than better off households. In the area where SEWA has introduced its wasteland development project, the poor collect over 70 per cent of their fuel and nearly 60 per cent of their fodder requirements from grasslands and fallows, whereas other households collect just over 50 per cent of their fuel and 25 per cent of their fodder from these sources (Chen, 1991).

Second, women are the primary gatherers and managers of biomass goods in poor rural households. As mentioned above, women perform key roles not only in the gathering but also in the processing, storing, utilisation, and marketing of free biomass goods. And, third, women's roles and responsibilities are pivotal not only to the management of natural resources but also to the management of the domestic economy. Recent studies have shown that women work longer hours, pool more of their income to household budgets, manage the day-to-day consumption and cash flow needs (often in the absence of men who migrate for better employment opportunities). Male migration is typically high in areas where natural resources have been most degraded (in the hills or on the fringe of degraded forests). In such areas, women are the more permanent residents whereas men are only intermittent residents.[11]

Poor rural households not only use a variety of biomass goods but also use them in a variety of ways, both for consumption and for sale. Some biomass goods are consumed directly, notably, most sources of fuelwood and fodder. Others are processed before being consumed or sold: some sources of fuel or fodder need to be cut/chopped, dried or combined with other products before consumption or sale (e.g., twigs are sometimes coated with animal dung and dried as fuel); some biomass goods are composted for manure; other biomass goods are processed as rope, mats, baskets, etc. Women are the primary processors, driers, and storers of many of these products. Moreover, women generally manage the energy flows from biomass resources, particularly in the form of fodder and composting materials, to the agriculture and livestock sub-systems of household livelihood systems: for instance, gathering and processing fodder,

[11] This fact has implication for the design and implementation of wasteland development schemes, as will be discussed later in the paper.

caring for animals, converting animal dung into fuel or fertiliser.

Women, Resources, and the Market Economy

Preoccupation with the impact of the fuelwood and fodder crises on women as actors in the domestic subsistence economy has obscured their equally important resource-based roles in the market economy. In addition to headloading (collecting wood for sale), which has received some systematic attention, women are actively engaged in wage employment for the Forest Departments or in resource-based industries and self-employment from collecting, processing and/or selling non-timber forest products.

It is estimated that two to three million people in India are engaged in headloading wood, most of them women (CSE, 1985; Agarwal, 1987: 181).[12] One study estimated that in the state of Rajasthan alone 400,000 households are engaged in the extraction of firewood from forests (NWDB, 1988: 15). Headloading is generally a low-return and high-risk occupation, as gathering of wood from reserved forests for sale is illegal.[13] Another study describes the working conditions of headloading as follows:

> Every day some 300 women firewood pickers disappear into the forests. They cut timber and greenwood, which is illegal. Sixty-eight per cent of them have been hurt either by the axe or by wild animals while collecting wood. They earn around Rs. 120 a month, and half of them are always in debt. They have a two-day cycle, walking as much as 12 km to collect fuelwood and then travel by train to the town for sale—along the way others make money off them; the railway man who allows them free on trains, the village head man who takes a cut, and the forest guard who looks the other way when forests are being axed. (Ninan, 1981)

Other forest products, notably bamboo, are also headloaded.

[12] This estimate refers only to commercial headloading and is exclusive of the daily loads women carry for subsistence needs.

[13] Collecting fuelwood for own consumption from protected forests is permitted in principle but looked down upon in practice.

Women also derive cash incomes through either wage labour or self-employment in the forestry sector. Wage employment can be of three broad kinds; regular wage labour for the Forest Department or Corporation; periodic wage labour under various social forestry programmes; and regular wage labour in resource-based enterprises (Kaur, 1988: 48ff). Both men and women are engaged in a variety of forestry-related tasks for the Forest Departments, including: nursery work; planting and tending of tree plantations; couping or marking trees which are to be felled; felling and logging timber; stacking timber; digging forest roads; collecting and processing NTFPs (Chand and Bezboruah, 1980: 70). While men carry out most of the felling, logging, and stacking of timber, women carry out most of the nursery work (bed-making, sowing or seedling preparation, weeding and transplanting). All other tasks are carried out by both men and women. Estimates of the total wage employment for women in the collection of NTFP and other tasks vary; one estimate is 300 million women days (Pant, 1980).

However, women are often paid lower wages than men for similar work, are not paid regularly, and are subjected to harassment if they complain (CIDA, 1988). Large numbers of women are also engaged, either as piece-rate workers or wage labourers, in several forest-based industries: notably, *beedi*-rolling and match-making.

As is now well documented, poor people—particularly women—collect a wide range of non-wood forest products for sale. Most NTFPs come from forests and are collected by tribals, although some are collected from grasslands or agricultural fields by non-tribal populations.[14] NTFP include fodder and grasses; raw materials like bamboo, cane, grasses for artisan activities; leaves, gums, waxes, dyes and resins; and many forms of food, including nuts, wild fruit, honey, and game.

Non-timber forest products are collected both for consumption and sale and play an important role in the livelihood strategies of the poor, both tribal and non-tribal. For some households, notably tribals living on the fringe of standing forests, the collection of NTFP is a primary occupation. For other households, notably non-tribals living on the fringe of degraded forests or grasslands, the collection of NTFP is a secondary occupation. For either group,

[14] It is estimated that 70 per cent of NTFPs are collected in the five states where 65 per cent of the tribal population lives—Maharashtra, Madhya Pradesh, Bihar, Orissa, and Andhra Pradesh (Kaur, 1988: 38).

collection has seasonal rhythms, depending on when the NTFP are ready to be harvested or when other occupations are available. In one area, most of the collection, especially for grasses, flowers, seeds, gums, resins, and lac, is carried out during the lean agricultural months from March to July (Gupta and Guleria, 1982: 133). Bamboo, cane, and some wild fruit can be collected throughout the year.

Until NTFP were nationalised, tribals were free to collect and sell NTFP as they wanted. But the nationalisation of NTFP has had two types of effects on tribals (and non-tribals) whose livelihoods depend in large measure on NTFP (Saxena, 1987: 8). First, their legal rights or legitimised claims to collect NTFP, whether for consumption or sale, has come under question. Forest guards and officials now often challenge tribal rights to collect NTFP. Second, tribals now have to rely on either the Forest Departments, Forest Corporations, or official contractors for the sale and price of NTFP. Private trade in most NTFP goods is now illegal in most states, although some private trade persists.

Other problems associated with the nationalisation of NTFP include: reduced number of legal buyers, interrupted flow of goods, and delayed payments. Tribal women in Madhya Pradesh told the National Commission on Self-Employed Women and Women in the Informal Sector that often when they walk long distances to the Forest Corporation office to sell NTFP, they find the office closed and are forced to sell the NTFP they have collected to traders at only 20 per cent of the government price (GOI, 1988a: xxv). A study of the collection and marketing of *sal* seeds in two villages in Madhya Pradesh reported that the majority of collectors were women; that the women had to sell the seeds to government agents; and that the women were paid only 20 per cent of the price at which the seeds were sold to oil-mills by the government (Dasgupta, 1986: 44–49).

Although there are regional and caste variations, which should be taken into account for planning purposes, women are the primary processors of NTFP. Their work includes: cleaning, drying and storing *mahua* flowers (used to distil liquor); collecting and drying seeds; cutting and bundling *datun* reeds (used for cleaning teeth) and manufacturing plates from *sal* leaves (Kaur, 1988:40). Two important cash earners are *sal* seeds and *tendu* leaves. Women are the primary collectors of *sal* seeds, which are crushed to produce an edible oil,[15] and *tendu* leaves, which are used as the wrapper in *beedi* production. Women are also the primary rollers of *beedis*, hand-rolled cigarettes.[16]

Many women are also self-employed in several resource-based enterprises, including: sericulture, basket-making, mat-weaving, and animal husbandry. Fodder can be collected either for consumption or sale, and animal products (milk, meat, draft power) can be either consumed or sold. It is widely acknowledged that women in many rural areas not only collect fodder but also provide most of the care of animals. One study from Himachal Pradesh reports that women contributed 69 per cent of the time and effort that went into tending animals (Bhati and Singh, 1987).

In summary, natural resources provide both subsistence needs and cash incomes to rural households, particularly poor rural households. And natural resources are of critical importance to the women of poor households as they gather, process, and manage these resources to a greater degree than men or better-off women. Poor rural women perform multiple roles (in regard to natural resources) as collectors or gatherers, as users or managers, as entrepreneurs, and as wage labourers.

Women, Men, and Natural Resources

In most parts of rural India, women have a different perspective on natural resources than men, even men from their own households. Typically, women maintain closer physical ties with the natural resource-base. As the primary collectors and gatherers of biomass goods, women make frequent, if not daily, trips into forest or grassland areas. A summary of studies in nine areas of India indicates that women made daily trips in five areas, trips every other day in two areas, and trips every four days in two areas (Agarwal, 1986a: Table 5). On an average, each trip covered

[15] It is estimated that the employment generated for primary collectors of *sal* seeds is eight million person days, over a collection period of 45 days, in Bihar, Madhya Pradesh, and Orissa (Tewari, 1981: 29).

[16] It is estimated that 106.11 million person days of employment are generated each year in the collection of *tendu* leaves. And that 675 million person days are generated in the rolling of *beedis* (Tewari, 1981: 27).

five kilometres and took four hours. Another summary of studies in 11 areas of India indicates that women average just under three hours per day collecting fuel; less than one hour per day in forested areas or where animal dung was readily available, and as much as five hours per day in the semi-arid plains or deforested hills (Kaur, 1988: 26a, Table VII).

As a result of this close physical tie, women have in-depth knowledge and know-how about the properties and uses of various biomass species. When it comes to the knowledge of fuelwood species, for instance, women can differentiate between those which provide quick high heat, those which provide long-lasting, low heat, and those which smoke. Women are also familiar with a range of biomass species other than fuelwood. In a study of one village in a semi-arid area in the state of Gujarat, women helped identify 13 varieties of trees, four varieties of bushes and shrubs, and 18 varieties of grasses that grew on the borders of agricultural fields or on common pastures (Chen, 1991: Table 16). Interestingly, traditional midwives have a long tradition of using medicinal herbs and plants for dealing with women's health problems (Sarin and Khanna, 1990: 8). When it comes to the management of fuelwood species, successive generations of older women have trained younger women in the art of lopping or pollarding (Shiva, 1988: 66).

Given their familiarity with the use and properties of various species, women are widely recognised to be more conservationist than men and to more readily perceive the negative consequences of the degradation and destruction of forests and grasslands. Some analysts, notably Shiva (1988), take this perspective one step further and argue that women are 'naturally' or instinctively closer to the forest, that preservation is part of the feminine ethos. However, not all women are engaged in resource-based activities and not all women share the same perspectives on natural resources.

Whereas women are the primary collectors of biomass goods, it should be noted that men and children also engage in collection. A study of rural energy sources in five different areas of India reported that 57 men and 110 women from 125 households collected fuelwood. However, women contributed 66 per cent of the labour time in collection whereas men and children contributed only 33 per cent (Maiti and Dasgupta, 1986). Other studies report that the men engaged in collection are generally younger than 15 or older than 40 years of age (Mishra, 1989: 20).

Because men and women perform different roles in regard to natural resources, women's knowledge about the variety of biomass species and their respective uses is often qualitatively different from that of men. Further, they often prefer different species than men do. The pioneering Chipko movement was the first to demonstrate that, in some communities or habitats, women are more likely than men to prefer subsistence species over commercial species. In one study of common property resources in Rajasthan, when consulted about species to be planted on village pastures, women preferred supporting a large range of local wild grasses, trees, and herbs which they used as sources of fodder, fuel, and food. The majority of men, by contrast, preferred planting trees that had a market value (Brara, 1987).

Project experience has also illuminated three other factors in women's choice of species. First, women often prefer indigenous species, over exotic species, and are the local experts on indigenous plants and trees. Second, women often prefer multiple species in mixed plantations, over monoculture systems, as they traditionally gather and use multiple products from both private and common land. Third, and a related factor, women often prefer intermediary products over timber, because 'they are accessible on a continuous or seasonal basis' (Ford Foundation, 1989: 13).

However, it should be noted that women's (and men's) preference for certain species is not uniform across India. In some projects, both men and women have expressed interest in the same species: subsistence, commercial, or a mix of both. For instance, in the tribal belts of West Bengal and Orissa where local forests are still a steady source of fuel and fodder, both men and women prefer commercial species (notably fruit trees). In some areas, men and/or women express interest in certain kinds of small timber for local construction and implement manufacturing. An interest in small timber for local use is often, mistakenly, construed as an interest in commercial species.

Finally, it should be noted that women face special constraints in regard to land and natural resources beyond those that men face: notably, limited property and political rights. In regard to private land, women have limited legal rights or legitimised claims. Where women lack property rights to private land, forest products and

CPRs are often the only source of an independent income for poor women. However, traditional use rights to forests and CPRs are vested in households or male heads of households. At best, therefore, women exercise tenuous claims to products from forests and CPRs despite their role as primary collectors and managers of biomass. The tenuous nature of women's rights to land directly affects their economic status. To illustrate, among many communities the only property that women have sole rights in is livestock. For livestock to be a source of income, fodder availability is essential (Kaur, 1988: vi). To ensure a steady supply of fodder, women need to secure use rights either to private land for growing fodder crops or to forests and CPRs for collecting fodder (ibid.).[17]

Project experience suggests that local men often resist efforts to strengthen women's ties to land, private or common. In their case study of SARTHI in Gujarat, Sarin and Khanna found that making decisions 'concerning land is almost an exclusive preserve of men which makes wasteland development with women a particularly difficult task' (Sarin and Khanna, 1990: 6). They report a 'normal resistance to women's involvement with land-based activities' (ibid.: 2).

Further, in many parts of India, men operate in the public domain while women operate almost exclusively in the private domain. For instance, men often come together in traditional village meetings from which women are excluded. Among the Bhil tribals in Dungarpur district in Rajasthan, there is 'a strong surviving tradition of men getting together twice every month on the 11th day of the lunar cycle to discuss village affairs. Women are completely excluded from this traditional village forum' (Sarin and Sharma, 1990: 15). In part because of their lack of political voice, women often do not have equal access as men to various programmes implemented for the benefit of the rural poor.

Because they generally lack both property rights and political voice, special attention needs to be given to women in wasteland development efforts. Although they are widely acknowledged to be the worst sufferers of environmental degradation, it cannot be assumed that women will automatically be included, or much less

[17] Refer to Kishwar (1987) for an analysis of how the absence of land rights affects the labour and economic status of one group of tribal women.

adopt leadership roles, in afforestation or wasteland development projects. Special outside attention is needed to ensure women a local role. For instance, many projects report the need to create a separate forum for women in which they can express their views, knowledge, and preferences uninhibited by the presence of men (ibid.: 15).

Impact of Ecological Degradation on Women

The well-publicised picture of the effects of environmental degradation on women, with which we are all now familiar, is that of women walking ever farther in search of fuel and fodder. A more complete picture includes the effects of shortages of biomass goods, other than fuel and fodder, on the time, income, and health of women and on the livelihood systems and family structures of the households in which they live. But these effects often remain hidden.

Time effects

Most studies of the impact of environmental degradation on women report a substantial increase in the time and energy spent collecting fuelwood and fodder (Agarwal, 1986a: 21; Molnar and Schreiber, 1989: 5, Box 5). The increased time and effort spent collecting water, especially during the dry season due to lowered water-tables or other factors, has been noted in some studies (Agarwal, 1986a: 24) . And the increased time and energy spent in collecting NTFP, other than fuelwood and fodder, has also been noted.

But environmental degradation has effects on how women use their time other than in collection, most directly on the time women spend cooking. Women have to spend more time on cooking either because they have to adapt their cooking methods to economise on fuelwood (Agarwal, 1986a: 21) or because they have to use fuel other than fuelwood, such as shrubs and crop residues which require more stoking and tending in return for less heat (Maiti and Dasgupta, 1986: 103). One study found that, on average, women spend 2.28 hours per day per household in cooking alone (ASTRA, 1981).

In addition, environmental degradation often forces women to spend less time on other activities. A detailed study from Nepal reports that women spend less time on agriculture in heavily deforested areas than they do in less deforested areas because of the time spent in collecting fuel and fodder (Kumar and Hotchkiss, 1988).

Income and livelihood effects

When women spend more time collecting biomass goods, they do not necessarily collect more goods. In fact, they are probably collecting less, which leads to decreased consumption or sale of these goods. And, as noted, more time in collection often means less time for other activities, such as agriculture, which results in lower levels of income or subsistence. Also, increased time spent in collection can lead to changes in the composition of income and livelihoods.

Nutrition and health effects

Lowered income and livelihood levels often lead to lower levels of consumption. Moreover, shortages of fuelwood lead to shifts in consumption patterns. Women adjust to fuelwood shortages either by reducing the number of meals cooked, shifting to less nutritious foods, shifting to foods which can be eaten raw, or serving partially cooked food or cold leftovers (Agarwal, 1986a). Such shifts in consumption patterns can lead to lower levels of nutrition in the household. The nutrition impact can be assumed to be particularly acute for women, who have to spend more time and energy in collection while eating less, or less well. The accumulated toll on women's health of more time and energy spent in return for less food or less nutritious foods needs to be studied.

Other health effects on women due to environmental degradation have been reported. There are cases where women have died or been injured while collecting fuelwood and fodder by falling from trees, falling down slopes, being cut by an axe, or attacked by wild animals (Dioman, 1989).

Effects on family structure

Environmental degradation has forced increased male out-migration in many areas, particularly in the hills and on the fringe of degraded forests. For significant periods of the year in these areas, women are the de facto heads of households, managing the day-to-day needs of children, themselves, and elderly dependents. In such areas, certain long-term demographic effects due to changed marital and fertility patterns are likely (Caldwell et al., 1988).

Impact of Afforestation on Women

Environmental degradation is not the only process that deprives women of natural resources. Ironically, afforestation can have some of the same negative effects on women as deforestation has. Under many afforestation projects, benefits are both gained and lost. Afforestation projects that fail to target women or to recognise the relationship of women to natural resources can have negative effects on women (Molnar and Schreiber, 1989: 11–12). At best, women may fail to receive or benefit from the expected project outcomes. At worst, women may also have to forego benefits which they enjoyed before the project was initiated.

In afforestation, what women gain and what they lose is affected by the choice of species, by the choice of land to be developed, and by the management regime introduced. The benefits women have foregone include grazing rights, foraging rights, legitimised claims to the land itself or to products from the land, and income or control over income. Often, species are chosen primarily for men's (cash-generating) enterprises and exacerbate shortages of products required by women for either subsistence needs or market-oriented activities (Molnar and Schreiber, 1989:12, Box 8). Further, women rarely have control over income from trees or other commercialised species (Kaur, 1988: 53).

Projects also often choose land units and management regimes which deny women the rights to grazing or foraging. Women's rights to graze or forage can be denied through conversion of common lands into woodlots to which access is restricted or closed; monoculture plantations on private or common lands which decrease the variety of available biomass goods and commercialisation of trees which effectively denies access to private farms for fodder and lops or tops (ibid.).

In addition, wasteland development projects which are not planned around women's schedules or needs may add to their workload. If the land brought under plantation is that from which they previously obtained fuel, fodder and other products and if the produce from plantations is not channelled to the women who need it, women may have to walk longer distances to gather fuel, fodder, and other products. Further, if afforestation activities are not planned around women's work schedules, women may have to spend unremunerated time in project activities plus additional time in routine collection activities.

Different Groups of Women

The well-publicised picture of women as the primary collectors of fuel and fodder obscures reality in three significant ways. First, as mentioned above, men and children also collect fuel and fodder, although to a lesser degree.[18] Second, as discussed above, women do more than collect fuel and fodder. They manage, process and protect a range of biomass goods and provide a large share of the labour force in forest-based activities and enterprises. Third, different women do different things. Indeed, some women do not collect fuel and fodder: notably, of course, most urban women but also some rural women.

For the purposes of this paper, women have been classified into four groups based on different habitats or agro-ecological zones (and related livelihood systems): plains women who live in the arid or semi-arid plains near degraded grasslands or forests; hill women who live in the hills near degraded or semi-degraded forests; tribal women who live on the fringe of standing forests and tribal women who live on the fringe of degraded forests.[19] In doing so, it

[18] Statistics reveal that women in the 25–40 age group are major collectors of fuelwood and fodder. Males who help are those below 15 or above 40 years of age: that is, those that are not active in the productive labour force (Mishra, 1989: 20).

[19] Kaur (1988) uses a similar typology but does not distinguish between the two types of tribal habitats.

is assumed that there are linkages between the ecological characteristics of any given habitat, local competition between different uses and users of natural resources, and the respective roles of men and women (Maiti and Dasgupta, 1986: 8).[20]

Most critically, local availability and utilisation of different sources of fuel and fodder can vary across habitats. For example, a study of 125 households in five different areas of India reported that in the areas near standing forests, fuelwood contributed 90 per cent of domestic energy, whereas in non-forest agricultural areas animal dung contributed 75 per cent of domestic energy (Maiti and Dasgupta, 1986: 8).[21] These variations can, in turn, affect women's (and men's) use of time. A summary of studies in 11 areas of India, as noted above, indicated that women average just under three hours per day collecting fuel: less than one hour per day in forested areas or where animal dung was readily available, as much as five hours per day in the semi-arid plains or deforested hills (Kaur, 1988: 26a, Table VII).

Tribal women on the fringe of standing forests

> The fuel need is fulfilled from the forest. I generally use firewood for cooking, heating water, and also for food preservation. I rarely use twigs or leaves. We also get material for housing from the forest. But sometimes we are harassed. The forest has been denuded in recent years. But new plants have also come up. The timber wood is removed by contractors. But they do not take fuelwood (Rabha Tribal Woman, Khasia Mountains, Assam).[22]

[20] Of course, there are variations within habitats. Carr-Harris (1990), for example, in her case study of Jakeshwar Sikshan Sansthan in Dasholi block of the Garhwal hills in Uttar Pradesh draws a distinction between a village which is relatively more resource-dependent (dependent on the local natural resource-base) and one which is more remittance-dependent (dependent on remittances from male migrants). This paper focuses on comparisons across habitats, rather than within habitats.

[21] This study reported that, on average across the five areas, fuelwood constituted 53 per cent of domestic energy, animal dung 24 per cent, leaves and twigs 4.4 per cent, corn cobs 5 per cent, and other sources 2.6 per cent.

[22] From Maiti and Dasgupta (1986: 29–30).

For tribals living on the fringe of standing forests, traditional use of the forests is still possible at least for the near future: notably, shifting agriculture and collection of NTFP.[23] Such communities still experience little difficulty in gathering sufficient fuelwood and fodder. Competition for forest resources comes largely from the Forest Department or private contractors and partly from non-tribal immigrants. Women play active roles not only in the collection of NTFP but also in shifting agriculture. Studies of the Ho tribe of Singhbhum district in Bihar (Kishwar, 1987) and other tribes in Orissa (Fernandes and Menon, 1987) report a greater role for women than men in shifting cultivation. And women are engaged in collecting NTFP not only as self-employed workers but also as wage labourers for the Forest Department (as discussed above).

Tribal women on the fringe of degraded forests

> Often we go to the forest to collect fuel. Many village women go to the forest in groups. These trips are strenuous. But we enjoy them. We need fuel for cooking, heating water, and also for distillation of beverages. The behaviour of the forest guards is not good. But still we must go. There are very few fruit trees now. All the *chikku* and *jackfruit* trees were cut down. In season, we get some mangoes and some berries. The timber trees have also been removed by the forest contractors. Some fuel trees are there. (Malhar Koli Tribal Woman, Thane district, Maharashtra)[24]

For tribals living on the fringe of degraded forests, traditional use does not still obtain. Because they can no longer derive traditional livelihoods from the forests, such communities are forced to collect and sell whatever is left of the natural resource-base. In such areas, headloading of firewood and/or bamboo for sale is common. One study of 170 households in nine villages of Ranchi district in Bihar State reported that headloading nad emerged

[23] It is estimated that 70 per cent of NTFP are collected in the five states where 65 per cent of the tribal population lives (Maharashtra, Madhya Pradesh, Bihar, Orissa, and Andhra Pradesh) (Kaur, 1988: 38).

[24] From Maiti and Dasgupta (1986:46).

as an important occupation in the previous 15 years (Agarwal and Narain, 1985a: 189). More than 20 per cent of the households in the study villages reported headloading as their major occupation. In most degraded forest areas, women are the primary headloaders. In addition, such communities often engage in settled agriculture on marginal lands and animal husbandry. In such areas, early competition with the Forest Department led to the current degradation of the forests. Current competition is largely between different local user groups over foraging and grazing rights and with non-tribal immigrants over agricultural land.

Plains women in arid or semi-arid plains

For people living in the arid or semi-arid plains, common grasslands, however degraded, are a source of products for agriculture (e.g., composting materials), livestock-rearing (e.g., fodder), the domestic economy (e.g., fuelwood and construction materials), and for sale (either processed or unprocessed). Here, the competition is largely between different local user groups over grazing and foraging rights.

Because the grasslands are often severely degraded, they are no longer a reliable source of fuelwood and fodder. Better-off households often shift to cultivating fodder and purchasing fuelwood. But poorer households, which do not have this option, often shift to using other biomass fuels (shrubs, agricultural residues, and animal dung) and to inferior sources of fodder (field stubble, agricultural residues, dry leaves and shrubs). A study of one village in a semi-arid area of Gujarat, reported that the poor collect over 70 per cent of their fuel needs and nearly 60 per cent of their fodder requirements from common grasslands or private fallows, while better-off households collect just over 50 per cent and 25 per cent respectively (Chen, 1991).

Hill women in the Himalayan foothills

For people living in the hills, where there are large tracts of semi-degraded forests, those forests are a source of inputs for local terraced agriculture and livestock-rearing.[25] Because of high

[25] Hill regions constitute 16 per cent of the land mass and eight per cent of the population in India.

levels of male out-migration from the hills, hill women carry out most of the local agriculture and animal husbandry. A study in one area of Himachal Pradesh reported that women constituted 62 per cent of the agricultural labour force (Bhati and Singh, 1987). In addition, women work long hours and walk long distances to collect fuelwood and fodder. In the hills, the competition over natural resources is largely between local users and outside commercial interests or the Forest Department.

Differences within habitats

Within each habitat, women can be differentiated by class, by caste, by household head, and by the degree to which they use CPR products (primary versus secondary users).[26] To illustrate, tribal women living on the fringe of a productive forest are more likely to prefer commercial species for cash income than plains women living next to degraded grasslands whose major concern is fuel and fodder. In areas of high male out-migration, women are the permanent residents and manage natural resources virtually single-handedly. Women from poorer or primary user households are more likely to be directly involved in collecting biomass than women from richer or secondary user households (which often either produce, buy, or hire labourers to collect their supplies of fuel and fodder). A female-headed household is more likely to be represented by a woman than a male-headed household if representation in local institutions is on the basis of one adult representative per household. The differences between women need to be understood and addressed in the design of wasteland development projects.

Section II: Wasteland Development

Five Broad Approaches

In the early 1980s, when the wasteland development movement

[26] Primary user households are those households which depend on biomass resources as their primary source of fuel, fodder, and other basic needs. The distinction between primary user households and secondary user households, those which depend less on biomass resources, is useful conceptually, although often difficult to measure.

had first started, the solutions appeared easy: identify a plot of land and mobilise people to plant some trees. Now, in the early 1990s, the solutions appear more complex. Wasteland development planners no longer talk simply about developing plots of land, planting trees, and generating income or subsistence. They discuss protection and plantation schemes to establish tree, grass, or bush cover on degraded grazing and forest lands; stabilisation of sand dunes and watersheds by establishing vegetative cover; agro-forestry or tree crop production on fields and field boundaries and improved management of common property resources.

This variety of project objectives notwithstanding, there are five broad approaches to wasteland development:

—Private management of private lands
—Private management of common lands
—Group management of common lands
—State management of common lands
—Joint management of common lands

Private management of private lands

There were two main components of early social forestry schemes: village woodlots and farm forestry. The targets for farm forestry were far exceeded, whereas there was a shortfall in village woodlots.[27] Farm forestry involves tree plantation or agro-forestry on small blocks of private land (fields or fallows) or on strips of private land (farm boundaries, internal field borders, roads, watercourses). Such schemes are designed to suit the labour and land resources available to individual households (Cernea, 1989: 46). Agro-forestry or tree plantation on private land can have at least two negative consequences: an implicit bias towards households which own land and a drop in demand for labour, as tree crops are less labour-intensive than annual crops.

[27] 'It is a well-established fact that of the various legal and institutional models of forestry in India farm forestry on private lands has shown best result' (Saxena, 1987: 48).

Whereas governmental wasteland development projects have tended to favour the individual household approach, non-governmental projects have tended to favour a group approach. In part this is due to the fact that most NGOs target the near-landless or landless, who have little or no land for farm forestry. In part this is also due to the fact that most NGOs favour group approaches in all their activities. Some NGOs have developed an alternative to farm forestry on private land by organising groups of households which 'pool' their private land in one or another way. For instance, in Rajasthan, Seva Mandir has encouraged neighbouring households to pool contiguous plots of private wasteland which they then develop in partnership. And, in West Bengal, the Centre for Women's Development Studies (CWDS) encouraged the men of neighbouring households to donate contiguous plots of private wasteland to a local women's organisation which then developed the land.

Private management of common lands

Various state governments have developed programmes to extend temporary ownership or usufruct rights to revenue wastelands (and occasionally degraded forest lands) to individual households or groups of households.[28] The basic approach is to extend ownership on usufruct rights to selected households which will, thereby, have a vested interest in developing that land. There are several such afforestation schemes in India including:

—Tree Patta Scheme: usufruct rights to trees grown on revenue land by individual households, targeted to Scheduled Castes and Tribes and to women;[29]
—Social Security Scheme: usufruct rights to forest land plus all inputs and monthly stipends to poor households for afforestation; intermediate harvest to poor; 20 per cent of profit after final harvest (15 years) to poor and 80 per cent to government;

[28] Refer to ILO (1988a: 52, Figure 1), for a summary of the characteristics of different social forestry schemes.
[29] As formulated, the Tree Patta Scheme 'distinguishes between tree tenure and land tenure: that is, the beneficiaries have no rights on land, their rights are confined to the usufruct of trees' (Saxena, 1987: 28: NABARD, 1989: 2ff).

—Farm Forestry for Rural Poor: usufruct rights to degraded government lands plus all inputs and daily wages (for the first three years) to poor households for both afforestation and agriculture;
—West Bengal Group Farm Forestry: ownership rights to contiguous plots of revenue land to groups of poor landless families for afforestation;
—Strip Plantations: usufruct rights to trees planted on the sides of roads, canals, and railways to individual households.

The West Bengal Social Forestry Programme includes a component called 'group farm forestry'. Under this component, contiguous plots of revenue land are granted to groups of marginal farm households or landless households which are encouraged to undertake block tree planting with shared labour and collective protection systems (Molnar, 1986a). The ownership of the land, the obligation to manage the land, and the right to dispose of the products are vested in the individual household. However, there is not only legal, but also group, control over the use and disposal of the land (Cernea, 1989: 57). In the other private leasing schemes, individual households operate the land autonomously.

A parallel NGO approach has been to negotiate rights to common land (or produce from that land) for targeted subgroups. Recently, these early efforts, both governmental and non-governmental, have been criticised as 'privatisation' of the commons because they extend exclusive rights to a sub-group over resources on which a traditional user group was dependent (Singh, 1990: 21). In describing an effort to involve landless poor in a 250-acre fodder plantation, one NGO leader explained:

> The community fodder farm was seen as privatising the commons. When the village did not have water, the farm had a bore well. Although there were 250 acres of *gomaal* or revenue grazing land available to the village, now there is hardly any left. Our hunch is that people started encroaching soon after we got this 250 acres, thinking that if they do not take it, the government will give it away to other special groups. As a result, there is hardly any *gomaal* land left for free grazing (Burra, 1990c: 15).

Group management of common lands

As mentioned above, one component of early social forestry projects was the establishment of 'community' woodlots on common lands by involving villagers in tree planting/protection and in sharing the benefits. Foresters and planners assumed that local village 'communities' would mobilise their members to collectively plant and protect 'their' land (Cernea, 1989: 33). While most social forestry projects surpassed their targets in farm forestry on private lands, none came close to their targets for establishing village woodlots. In some states, management of the village woodlots was then handed over to the village *panchayats* (or the state forestry departments). Evaluations of various social forestry projects indicate that, in many cases, the villagers were not even aware that the woodlot had been established for the village because efforts to organise the 'community' had not been undertaken (ibid.: 35ff).

On the basis of this early experience, some foresters and planners argue that, without prior organising and without active participation in the design and implementation of a scheme, village 'communities' cannot collectively plant and protect 'their own' woodlots. In response, the recent governmental approach has been to organise local cooperatives or groups to avail of the alternative social forestry schemes listed above: especially, the Tree Patta Scheme. The National Dairy Development Board, for example, has organised a large number of Tree Growers' Cooperatives (Mishra, 1989).

Many environmentalists and NGOs take the issue of group management one step further, arguing that the real issue is not just whether local people are organised but also who is organised. They recommend identifying and organising those groups which had traditionally used or managed the common land. Most 'village' land, they point out, has traditionally been managed by user groups that are smaller (and more homogeneous) than village *panchayats*.[30] In response, many NGOs have shifted from the earlier 'privatisation' approach (i.e., leasing land for targeted sub-groups) to surveying existing

[30] As mentioned above, a village often consists of several settlements/hamlets and a *panchayat* usually covers several villages.

land use systems to determine which groups traditionally used which units of land and to organising those groups to regain access and management rights to that land.

Government management of common lands

The Government Forest Acts of 1865 and subsequent legislation have gradually eroded the traditional use rights and management systems of local user groups, undermining their ability to control forest and common land use. Experts differ on the degree and nature of traditional common property resource management. One forest law specialist estimates that, prior to this legislation, 'at least 80 per cent of India's natural resources were common property' (Singh, 1986:10). Whereas now, according to two environmental specialists, nearly one-third of India's land and all its water resources are owned by the government (Agarwal and Narain, 1989: 27).

The right and responsibility of the government to manage sections of forest lands for the public good, including the need to meet the material needs of industry, are acknowledged by some environmentalists. However, government management of village or revenue lands, without involvement of local user groups, is questioned by most environmentalists. Under the early social forestry programme, the Forest Departments repeatedly established 'community' woodlots on village or revenue grasslands without consulting the local people.

In such cases, the villagers assumed they were 'government woodlots'. For example, in one village in Gujarat, two 10-acre plots of revenue land have been developed as village woodlots. The villagers were not consulted either about the decision to convert what had been open grazing land into protected woodlots or about the species to be planted. The only local involvement was short-term wage employment for a few villagers in establishing the plantations and part-time salaried work for two villagers in protecting the plantation. Because the trees have not as yet matured and benefits have not as yet been distributed, the villagers assume the benefits will go to the government (Chen, 1991).

Joint management of common lands

In West Bengal, the forestry department has developed an

innovative and effective approach to involving villages in joint management of degraded forest land. This approach, earlier called the Arabari project (after the forest range in which it was first tested) and now referred to as the Forest Protection Committee (FPC) scheme, organises villages or settlements on the fringe of forest areas to protect (by controlling access) adjacent blocks of forest land. Under a verbal agreement with the Forest Department, in return for protecting this land for a minimum of five years, the FPCs, consisting of one representative from each household, are entitled to 100 per cent share of intermediate yields from non-timber forest products and a 25 per cent share of the final timber harvest to be distributed in cash (the local *sal* trees are clear felled every 12–15 years).[31] Also, FPC members are guaranteed employment in any local forest management operation (e.g., cutting secondary shoots from degraded *sal* stands). By the end of 1989, about 1300 Forest Protection Committees had protected and, thereby, regenerated 152,000 hectares of degraded *sal* forest (Viegas and Menon, 1990).[32]

The State Governments in Gujarat and Orissa have recently passed resolutions to introduce the Forest Protection Committee approach in their respective states (Sarin and Khanna, 1990; Viegas and Menon, 1990).

People's Participation

Before analysing how women have been (or can be) involved in wasteland development, it is important to understand precisely

[31] In early 1990, the Government of West Bengal's Forest Department, in consultation with the FPCs and social science advisors, was working out the details for written legal agreements to be signed by the Forest Department and each FPC.

[32] The Forest Protection Committee scheme has been implemented in Bankura, Midnapore, and Purulia districts of West Bengal. Similar efforts at forest protection by adjacent villages, at the initiative of NGOs or the local communities themselves, have been reported in Orissa and Madhya Pradesh (P.V. Rajgopalan, Federation of Voluntary Agencies). And the well-known Chipko movement started as a spontaneous effort, largely by women, to protect local forests from commercial felling.

how people in general have been involved. Most early social forestry or wasteland development projects involved people largely as a way to mobilise the necessary labour force. Many current projects, both government and NGO, involve people in an attempt to meet both distributive and environmental objectives. The reasons why wasteland development projects choose to involve local people can be grouped broadly as follows:

Economic objectives

—to provide labour for afforestation on a large scale;[33]
—to increase production and supply of timber, wood products, and NTFP;

Distributive objectives

—to transfer real resources to a designated group;
—to produce basic biomass needs for a designated group;

Environmental objectives

—to enhance total biomass base of a given ecosystem;
—to create viable local institutions to manage common resources on a sustainable basis.

What follows is a framework for analysing whether and how people have actually been involved in wasteland development. The headings represent the various stages of wasteland development. And the comments under each heading describe the various modes of involving people: for example, as labourers, as legitimate clients, or as full partners (with ownership and management rights).

Pre-project activities

—little, if any, prior contact with people;

[33] Systematic and effective afforestation is seen to require more or less the same size of labour force as agriculture (Singh, 1990: 1).

—prior organising of people around other activities;
—organising of people in preparation for afforestation activities.

Project planning and design

—limited consultation with people regarding species and site selection;
—consultation with people regarding species and site selection;
—involvement of people in deciding which species and sites;
—consultative micro-planning with people.

Methodologies for consultative micro-planning are being tested in various parts of India: in watershed development (Karnataka and Haryana) and forest management (West Bengal, Gujarat, and Orissa). The common elements in these experiments include: rapid rural appraisal of local site conditions from a local perspective; collection of information about present patterns of land use and tenure status; sketch mapping to determine locations and boundaries of different land resources; community profiles to determine key social, economic, and ecological characteristics of each participating group; and group discussion techniques to allow community members to identify specific development needs, opportunities, and priorities (Ford Foundation, 1989: 44).

Organising and building local institutions

As mentioned above, the central question in organising local institutions to manage common resources is the question of who is to be organised. Most government schemes aim to organise people either at the level of village *panchayats* which often cover more than one village (as in the early community woodlots scheme) or at the level of individual villages (as in the Tree Growers' Cooperatives or the Forest Protection Committees). Many NGOs aim to organise smaller, more homogeneous units, such as traditional user groups. Experience suggests that the success of group management varies in inverse proportion to the size and heterogeneity of the management group (Arnold et al., 1989).

The next question in organising local institutions is the question of to what degree the organising involves mere 'tokenism' or real involvement in the design, implementation, and management of schemes. The issue is complex because local institutions clearly have to perform two vital functions. They are expected to represent, in a fair and equitable way, the particular group being organised. They are also expected to assume responsibility for the long-term management of the scheme (ibid.: 20). The framework presented here is intended to help analyse whether specific approaches to people's participation will lead to either (or both) of these goals: that is, to equitable representation of the people being organised and/or to the efficient management of the scheme being undertaken.

Other than as wage labourers, people's participation has generally to be mediated through one or another form of representation. The institutional alternatives to representation include: direct representation (of all members at open public fora), designated representatives from each household (usually one adult per household), elected representatives, or assumed representation (by traditional leaders).

Negotiating land rights

—limited, if any, consultation with people in allotting common land;
—allotment of common land by government decree rather than through negotiated settlement with people;
—consultation with people in negotiation between the NGO and the government;
—allotment of land in the name of the NGO;
—allotment of land in the name of the local institution.

Seed collection and nursery work

—hiring of people as wage labourers to work on nurseries;
—supporting people to develop their own nurseries as a source of self-employment;
—organising or paying people to collect local varieties of seeds.

In the 1970s, most seedlings were raised in nurseries operated by the Forest Departments. But as afforestation efforts expanded, these government nurseries could not meet the demand for seedlings.[34] In 1986, the National Wasteland Development Board introduced a large-scale Decentralised People's Nursery Scheme to decentralise production of seedlings by transferring nursery technology to local farmers, schools, and specified disadvantaged groups—landless, Scheduled Castes and Tribes, and women—(NABARD, 1989). As part of this shift to decentralise nurseries, local men and women were organised and paid to collect local indigenous varieties of seeds.

Land preparation

—paying people's wages from project funds to clear/level land and dig pits;
—mobilising voluntary labour to clear/level land and dig pits.

Plantations

—paying people wages from project funds to plant saplings;
—mobilising voluntary labour to plant saplings.

Protection

—paying people wages from project funds to build boundaries or to protect plantations;
—organising people to contribute to group funds which are used, in part, to hire guards;
—organising people to develop local systems for protecting plantations through voluntary labour, usually on a rotation basis.

Maintenance

—paying people wages from project funds to clean, weed, or prune plantations;
—organising people to contribute to group funds which are used, in part, to pay people to maintain their plantations;

[34] To achieve the National Wasteland Development Board's annual afforestation target of five million hectares, it was estimated that 10,000 billion seedlings would be required (Krishnan, 1989).

—organising people to develop local systems for maintaining plantations through voluntary labour, usually on a rotation basis.

Collection and harvesting

In terms of collection and harvesting, it is important to distinguish between intermediate harvesting on a continuous or seasonal basis of grasses, dead wood, bamboo, shrubs, and NTFP (typically carried out by women) and the one-time final harvest of small timber and timber (typically carried out by men):

—regulating harvesting of intermediate products by people;[35]
—organising people to develop local systems for equitable harvesting of intermediate products;[36]
—organising people to develop local systems for harvesting timber.

Processing

—processing by individual households using traditional technologies;
—supporting people in small-scale forest-based enterprises.

Marketing

As most afforestation projects have not yet reached maturity and most intermediate products (harvested during the maturation period) are consumed locally, few systems for marketing have yet been developed. Some projects have marketed intermediate products, as follows:

—marketing of intermediate products by individual households (through Forest Corporations, marketing cooperatives, Tribal Corporations);
—marketing of intermediate products by the local institution or the implementing agency (e.g., NGO).

[35] Regulation typically includes charging a permit fee to each household for collecting/harvesting and/or imposing fines on offenders.
[36] Including permit fees and penalty fines.

Before the final harvest of timber, a number of decisions will need to be made regarding who harvests, who the poles are sold to, who sells the poles, the pricing of the goods, and the distribution of the benefits (Ford Foundation, 1989: 16).

Benefit-sharing

Various systems operate for the distribution of intermediate products. But few systems for the distribution of small timber and timber have yet been developed, given that most plantations have not yet matured. Under many schemes it remains unclear who is to benefit from timber sale. However, a few schemes have worked out distribution systems.

In the case of a successful village woodlot scheme in Gujarat, the *panchayat* decided that the labourers who harvested the plantation could remove headloads of lops and tops and that landless, small and marginal farmers could buy fuelwood at a highly subsidised rate and poles for home repair at a rate of 60 per cent of the market value. The rest of the produce was sold by auction and the village *panchayat* received 35 per cent of the total revenue (ILO, 1988a: 57). By way of another exception, the Forest Protection Committee scheme in West Bengal has stipulated in its plans that the local Forest Protection Committees will receive a 25 per cent share of timber value distributed in cash.

Since a few plantations have been felled to date, there is little experience with sharing pole revenues. Procedures need to be developed to ensure that the benefits are not captured by village elites, corrupt officials, or middlemen (Ford Foundation, 1989: 16).

—Rationed collection by all households on a first come, first serve basis;
—free collection by all households on a first come, first serve basis;
—free collection by all households on a rotation basis;
—free collection by a targeted sub-group on a priority basis, followed by other households;
—free collection of twigs and leaves by those who prune or thin the plantations (Sarin and Sharma, 1990: 33).

Profit-sharing and investment

Again, because such a small number of projects have reached maturity, a few (if any) systems have been developed for sharing or investing profits from collective sales. By way of exception, the CWDS project in West Bengal deposits profits from the sale of silk cocoons at the end of each production cycle in a revolving fund, which is used to cover wage payments during the next production cycle.

Monitoring and evaluation

—monitoring and evaluation by government, international donor agencies, NGOs or other outsiders;
—participatory monitoring and evaluation.

Section III: Women and Wasteland Development

Few studies have addressed women's roles in either traditional or new management regimes, other than presenting reasons why women should be included. This Section addresses several key questions relating to women's participation in wasteland development. Why should women participate in wasteland development? What should be the nature of women's participation? Should they be treated as an exclusive unit for intervention, or on an equal basis alongside men, or should women's participation simply be subsumed under male or household participation?

Why Women should be Involved

Thinking about why women should be involved in wasteland development has gone through several phases. In the 1970s, women were first recognised as special victims of the environmental crisis: as those forced to spend increasing amounts of time and energy collecting fuel and fodder. To offset the cost of environmental degradation to women, women were seen as special targets of early social forestry projects. But these early projects involved women primarily as wage labourers in nursery or plantation work.

Since 1980, women have increasingly been recognised not only as the victims but also the managers of the environmental crisis: as those who help their respective households weather the crisis. Given their key roles and knowledge about natural resources, women's participation is now seen to make a 'crucial difference for ecological regeneration programmes' (Agarwal and Narain, 1989: 26). Indeed, some analysts argue that women's participation in afforestation projects ensures that projects maximise returns on investments and achieve their goals (Molnar and Schreiber, 1989: 12). Some projects, like those funded under the ILO's Women and Wasteland Development programme, conceptualise women's participation in empowerment terms. Such projects seek to promote the empowerment of women by transferring real resources to them and by increasing their participation in local institutions.

Currently, three sets of arguments are being made for why women should be involved in wasteland development: to meet project objectives, to meet welfare objectives, and to meet equity objectives.

Project objectives

There is a growing recognition that, unless women who are deeply affected by the environmental crisis are integrally involved in any effort to overcome the crisis, the effort will achieve only partial success. Because women depend more heavily than men on natural resources they generally perceive the implication of lost forests and grasslands sooner and better than their male counterparts (Molnar and Schreiber, 1989). For instance, women have taken dominant and leading roles in movements for the protection of forests (Kaur, 1988). Because of their 'enlightened self-interest' (Lekha, 1989: 8), women are often a stronger lobby for the protection of natural resources. As reported in the case study of PEDO in Rajasthan, 'once women get mobilised to act on a particular problem, they leave the men trailing behind in initiative and commitment, sometimes having to drag the men along' (Sarin and Sharma, 1990: 30).[37] Also, because women are knowledgeable about local species, their involvement snould help preserve species which might otherwise be lost through degradation or monoculture (Kaur, 1988: 5). Or involving women may simply be a necessity, because labour for afforestation is required on a large scale (Singh, 1990: 1).

Further, female-headed households (or households without adult males) may form a large percentage of households in any given project area. In India as a whole, about 10 per cent of households are reported to be headed by women. In areas of high male-outmigration due to environmental degradation, women become *de facto* household heads: particularly in the hills but also in the semi-arid plains or tribal belts where local forests have been degraded. When female-heads of households are encouraged to participate, project returns will most likely increase (Molnar and Schreiber, 1989).

If they do not perceive project activities to be in their own interests, women will not (or will only reluctantly) participate in wasteland development projects (ibid.: 12, Boxes 7 and 8). However, if women's species or product preferences and women's time schedules are taken into consideration, women will have more inclination and time to participate in project activities.

Wasteland development projects may even fail if women are not involved or involved only partially. For instance, if women are responsible for key project activities such as nursery work, tree planting and tending, but do not receive the extension messages and requisite technical know-how, seedlings may not survive and thrive (ibid.). Further, if forest produce from plantations is not channelled to the women who need it, the pressure on existing natural resources from women gathering wood may increase rather than decrease (ibid.: Box 8).

[37] Women's responsiveness and responsibility in other related programme areas have also been noted. For instance, rural women have been shown to be better loan repayers than men. Therefore, targeting women rather than, or in addition to, men for institutional credit for planting or land improvement may have higher returns (Molnar and Schreiber, 1989).

Welfare objectives

It is now widely acknowledged that women's income goes directly to meeting subsistence needs and raising household living standards. Wasteland development projects can generate goods for subsistence, direct cash incomes, or indirect cash incomes (e.g., if raw materials for home-based industries are available) (ibid.). To the degree that wasteland development efforts serve to increase women's independent income or women's control over project benefits, the welfare and living standards of their households should also increase.

Equity objectives

Most wasteland development projects which involve women aim to meet the subsistence needs of women and their families, to increase women's income, and/or to reduce women's burden and drudgery. These objectives satisfy many of women's practical needs. A few wasteland development projects also aim to address the more strategic needs of women, notably: their need for property rights and their need for political representation.[38]

'Where women lack property rights in land, forest products and common property resources (CPR) provide the only sources for an independent income for poor women' (Kaur, 1988: vi). Wasteland development projects typically involve the development of alternative tenurial systems over local resources and of local institutions to manage these resources. Therefore, wasteland development projects offer a unique opportunity to transfer real resources to women and to promote women's representation in local institutions.

How Women have been Involved

We, at the Tree Growers' Cooperatives Project of NDDB, have

[38] The useful distinction between practical gender needs and strategic gender needs was developed by Caroline Moser in a recent article (Moser, 1989). Moser's definition of practical and strategic gender needs will be discussed in the final section of this paper.

> recognised that (the) women's involvement means not just hiring women labourers, but ensuring that women participate in the planning, the execution, and in the sharing of benefits .... (V.K. Mishra, 1989) National Dairy Development Board.

As discussed above, there have been five broad approaches to wasteland development in the past. Women have been involved to varying degrees under each of these approaches. They have been least involved in schemes to promote private management of private lands, such as farm forestry. This is because women face the same constraints in farm forestry that they face more generally in agriculture. These are the lack of rights to land (as men generally own private land in India) and the lack of access to credit, inputs, and extension services which are typically directed to men as landowners. Women have also not been effectively involved in village woodlot schemes, proposed to be managed by village *panchayats* but generally managed by the government. When managed by local *panchayats*, women are effectively excluded as their representation in local *panchayats* is limited. When managed by the government, women, as well as people in general, are effectively excluded.

Efforts to involve women in private management of common lands by extending the use of ownership rights to women have been more successful. The Tree Patta Scheme of the National Wasteland Development Board was designed to confer assets to women by insisting that not less than 30 per cent of tree *pattas* should be issued in the name of the woman or as a joint *patta* in the names of the husband and wife (Krishnan, 1989: 19). In their background papers to a Workshop on Involvement of Women in Social Forestry (1989), both the NWDB and NABARD (which provides institutional finance to wasteland development projects) recommended expanded coverage of women under the Tree Patta Schemes (Krishnan, 1989; NABARD, 1989).

Efforts to involve women in group management of common lands have been quite successful: most notably, efforts by NGOs but also some efforts by the government. As the quote above suggests, the National Dairy Development Board has deliberately attempted to promote women's participation in their Tree Growers' Cooperatives by insisting that women attend all cooperative meetings, that 50 per cent of all households be represented by women, and that at least three of the nine management committee members elected are women (Mishra, 1989: 2–21).

Mechanisms to insure women's full and equitable participation in the joint management of common lands, specifically in the Forest Protection Committees, are currently being discussed. At present, in villages where FPCs have been established, each household nominates one representative to the committee. Most households, because of the traditional exclusion of women from public meetings, send male representatives to committee meetings. In one FPC, visited by Neera Burra of the ILO and myself, there were only five women members out of a total of 70 members: each of these women was a household head. In their case study of FPCs, Viegas and Menon (1990) discuss efforts to promote women's representation in the FPCs. It has been recommended that the contracts signed between FPCs and the Forest Department (which stipulate norms for benefit-sharing) be signed by both husband and wife in male-headed households and by women in female-headed households.

There are three general approaches in regard to women's participation in wasteland development: those that argue for exclusive participation, for joint participation, or for separate participation. The exclusive participation approach targets women as a discrete unit for project interventions: for example, the CWDS project in West Bengal and the afforestation efforts through *mahila mandals* in Himachal Pradesh. The joint participation approach targets project interventions to both men and women but does not organise women separately. The separate participation approach targets both men and women but recognises the need to organise women into separate groups, at least initially, to promote and protect women's interests. That is, women are seen to need a separate forum in which they can express their views and preferences uninhibited by the presence of men (Sarin and Sharma, 1990: 15).

In terms of how they involve women, wasteland development projects have involved women in much the same way as they have involved people (more generally) but with a few differences, as noted below:

Pre-project activities

> Under social forestry programme we have women motivators who are engaged in travelling from village to village and contacting people and motivating them for taking up tree planting. They are the vital link between departmental field staff and common people. Experiences have shown in many areas that they are more acceptable to the village women folk, as a result very satisfactory adoption response was witnessed in many areas. (Forest Department Official, West Bengal)[39]

Most wasteland development projects have effectively involved women engaged in prior organising and promotional activities in order to convince both men and women of the importance of involving women. Exposure activities to convince the women themselves included study tours to other projects, camps or fairs, training programmes, and other types of women-to-women exchanges. A notable example is the three-day women's *mela* organised by PEDO in southern Rajasthan in 1987, attended by 3000 women (including women from SARTHI in nearby Gujarat), which served as a catalyst to both the PEDO's and SARTHI's efforts to involve women in wasteland development (Sarin and Sharma, 1990). Similar exposure camps or fairs have been organised by many projects in the Himalayan foothills and by the Centre for Women's Development Studies in West Bengal, among other places.

Once women have been motivated, they prove very responsive (as discussed above). Some government projects have used women as motivators and as a link between local people and the Forest Departments: for instance, in West Bengal (as reported in the quote above) and in Himachal Pradesh (Dioman, 1989: 5).

Given the normal resistance to women's involvement in land-based activities, reported in the case studies of the PEDO and SARTHI, prior organising and motivating activities to convince men are critical. Joint meetings with both men and women to explain project objectives and to seek consensus through open discussion (on choices regarding membership and land) can be very effective (Sarin and Khanna, 1990: 2; Mishra, 1989).

[39] As reported by Das (1989).

Project planning and design

> Another area, where we attempt to involve women is in planning. During initial household surveys, since we were new and had not built up credibility, it was usually men who were the respondents to our questions. When asked about the scarcity of fuelwood and fodder, their replies were ambiguous, and they were more keen on growing fruit and timber trees, because this would bring them direct cash. They did not feel the real shortage of firewood .... it is only when we have established some credibility and have access to speak to women, during the preparation of village micro plans, that the true picture emerges. It is then that women become a part of the planning process. They decide what trees should be cultivated and what their primary needs are (Mishra, 1989: 21).

Women have an essential role to play in planning plantations and choosing species. Many projects have shown that women's choice of species is often different than men's. In their case study of PEDO, Sarin and Sharma (1990:10) describe a planning exercise in which women estimated 'the number of trees and shrubs a particular member of the group required for meeting her own firewood, fodder, construction material and other subsistence needs'. The planning exercise 'generated tremendous interest and excitement among the women and provided them a concrete basis for discussing action strategies with other women' (ibid.).[40] In Himachal Pradesh, one representative from each official *mahila mandal* serves as a member of the local Village Development Committee (VDC) which is actively involved in developing local plans for integrated resource management (Dioman, 1989: 3).

Organising and building local institutions

The consensus that emerged was that although women were the

[40] In this planning exercise, the women calculated that 'for supporting a cow or a buffalo entirely on tree leaf fodder for the whole year, 90 trees would be required for just one animal. For a goat, 50 trees would be needed' (Sarin and Sharma, 1990: 10). They compiled a list of 21 local species suitable for fodder. Similar calculations were made for the timber required for building a two-roomed house. The women were amazed at the number of trees needed just to meet their basic needs.

worst sufferers of environmental degradation, their lack of land ownership and traditional male domination effectively inhibited them from playing a leading role in plantation related decision-making, except in a few villages. Joint meetings of men and women, while successful in reducing male hostility and securing their cooperation, tended to diminish free expression and articulation by women. The need to create a separate forum for women in which they could express their views and concerns uninhibited by the presence of men, was strongly felt (PEDO Case Study).[41]

Wasteland development projects have the choice of organising women into exclusive all-women institutions, into joint insitutions with men, or separate women's institutions which maintain ties (or eventually merge) with men's institutions. Each approach to organising has implications for the degree of political representation women achieve:

Exclusive Representation

In exclusive all-women's local institutions, women's voice is most effectively heard. However, some analysts argue there is a potential liability in exclusive all-women projects. Since traditional use rights to common lands were vested in households or male-heads of households, involving only women in wasteland development could be interpreted as a violation of traditional rights and could lead to avoidable conflict.

Joint Representation

There are two ways in which women can be represented in joint men–women's local institutions, either indirectly (through nominated or elected male representatives) or directly (through nominated or elected female representatives). Most commonly, joint local institutions are constituted of one nominated representative per household. Under this arrangement, male representatives predominate as men easily assume public roles. Sometimes women representatives are mandated as members of local institutions. For example, one woman from each *mahila mandal* serves as a representative on the local Village Development Committee in Himachal Pradesh (Dioman, 1989). The

[41] From Sarin and Sharma (1990: 15).

> NDDB tries to ensure that at least 50 per cent of households under its Tree Growers' Cooperatives be represented by women and that three of the nine management committee members of each cooperative are women (Mishra, 1989: 21).

Separate Representation

> Many projects argue that since traditional rights to use common lands are vested in households, working exclusively with women would only generate avoidable conflicts. Yet, because women often cannot participate effectively in joint groups, separate groups for women which can eventually merge or be affiliated with men's groups should be organised. Many projects report that once women are able to get together regularly, to participate freely, and to have access to information on a regular basis, they gain 'the confidence to articulate their views even in meetings where men are present' (Sarin and Sharma, 1990: 30).

Three other facts about women's participation in local institutions should be noted. The first is that some projects initiate wasteland development projects with existing women's organisations, whereas other projects organise new women's organisations specifically for wasteland development. The second is that organising women into separate institutions does not preclude domination or manipulation by men. In her case study of Jakeshwar Sikshan Sansthan, Carr-Harris reports that men were taking the decisions in the name of the local *mahila mangal dal*. The women explained that they were 'unwilling to act contrary to their husbands and village decision-making bodies' (Carr-Harris, 1990: 24). At least initially, regular supervision and support from outsiders is needed to build the women's strength and confidence. Third, since they have generally been sidelined by traditional leadership structures and government services, women are less manipulative and corrupt in their institutional behaviour and more 'prone to honouring the commitments they made' (Sarin and Sharma, 1990: 30).

Negotiations for land rights

As with people's participation in wasteland development more generally, the central issue in women's participation is that of ownership and usufruct rights. How wasteland development projects approach the issue of women's tenurial rights is a key variable in distinguishing between them, reflecting the underlying philosophy of each project. [42] As with women's political representation, there have been three different approaches to women's tenurial rights:

Exclusive Rights

Under this approach, women are organised separately from men and the right to land/produce is negotiated in the name of individual women or their organisations. This is generally the rule with projects which are developed exclusively for women (e.g., CWDS in West Bengal) but it is also the case in some projects which work with both men and women (e.g., *mahila mandals* in Himachal Pradesh).

Joint Rights

The right to land/produce is negotiated in the names of both male and female members of a household or group. For example, under a forest concession scheme in Haryana, one man and one woman from each household in selected Banjara communities are joint signatories to contracts with the Forest Department which entitle them to collect *banjara* grass from the forest.[43]

Indirect Rights

The right to land/produce is negotiated in the name of the local institution which has been organised for the purpose of wasteland development. Often women are not equally represented in these local institutions and may not, therefore, enjoy equal benefits with men.

One important fact to be noted is that tenurial right, whether of men or women, are often not clearly negotiated and deeded. The ambiguous legal status of ownership or use rights to common land,

[42] Of course, different approaches to women's land rights are directly influenced by the type of land available and the specific state laws pertaining to that land.

[43] Personal communication, Mark Poffenberger. The negotiation of collection leases for the Banjara community, rather than to middlemen who sub-contracted to the villagers in the past, was part of the Hill Resource Protection Committee scheme, a community-oriented watershed management programme in the Shivalik Hills in Haryana (Ford Foundation, 1989).

whether under traditional or new management regimes, is a central problem to both men and women. Women face the additional problem of the ambiguous legal status of their rights to private land.[44] A central concern to many women's groups, therefore, is 'how to legally secure their rights to the produce of the land they are developing with so much effort' (Sarin and Khanna, 1990: 34).

Nursery work and seed collection

Many projects, both government and NGO, give preference to women as wage labourers in nursery work, as women are seen to be skilled at handling seedlings. Also many decentralised nursery projects, both government and NGO, train and support women to develop their own nurseries as a source of self-employment. One report on the decentralised nursery programme of the West Bengal government indicates that women benefited from 320,000 days of employment (out of a total of 800,000 employment days generated). (Das, 1989: 3)[45]

Some projects, such as PEDO, have promoted the collection and procurement of local seed varieties by women.

Land preparation

According to a NDDB official, during plantation activities it was evident for all to see that women were gentler, worked faster and more efficiently in planting saplings and also in the aftercare (Mishra, 1989: 21).

As observed in the PEDO case study, there was always a much higher percentage of women than men doing the earthwork and planting. Women's productivity was always higher than that of men (Sarin and Sharma, 1990: 19-20).

Women's 'participation' in official social forestry programmes often amounts only to wage employment in nursery work, land preparation, and plantation activities. An evaluation of the

[44] This problem is particularly acute in areas of high male out-migration, such as in the hills, where women manage natural resources singlehandedly for much of the year and yet feel very insecure about their rights to assets 'in case their link with the outside family member gets severed' (Carr-Harris, 1990: 31).

[45] Refer to Verma (1988) for an analysis of SEWA's decentralised nursery scheme for women.

government's social forestry programme in Orissa reports that men were engaged in 70 per cent and women 30 per cent of the more than 6.5 million work days generated over three years (Olsson, 1988). But as another evaluation argues, women's participation in wage employment under social forestry 'is not, apparently, the result of efforts to target them but rather because ... poor women are normally among the principal daily wage-earners in the countryside' (Arnold et al., 1989: 25). In fact, the 'female employment pattern in social forestry seems to follow the overall pattern of female participation in economic activities. It was quite low in Jammu and Kashmir presumably because of cultural attitudes. In Maharashtra and Gujarat, the proportion was quite high at 46 per cent (ILO, 1988a: 40).

Many non-government wasteland development projects report that, when manual work is carried out for wages, women work alongside men in equal or greater numbers, but that, when manual labour is carried out on a voluntary basis, women's participation is significantly higher than men's. These projects explain that women generally take greater interest and responsibility in the plantations. In their case study of PEDO, Sarin and Sharma (1990:18–19) report that although both men and women had immediate interest in getting daily wage employment, women were more willing than men to discuss long-term protection and management issues; to contribute a small amount every month to pay for a watchperson's salary; to self-monitor work output to ensure the minimum productivity required; and to take a stand against certain households which demand a disproportionate share of the employment generated on the plantation.

Women face special constraints in regard to wage work or manual work beyond those that men face: notably, differential wages and workloads. Many projects, both government and non-government, pay women lower wages than men (Olsson, 1988; Burra, 1990 a,b,c). The general excuse offered is that women receive lower wages in local labour markets. Women in many rural areas of India work longer hours each day than men. In one village in Gujarat, women work on the average 3.5 more hours per day than men (Chen, 1991). Therefore, women often find it hard to adjust their project-related work to their continuing need to fetch fuel, fodder, and water and to their routine work. Some wasteland development projects, such as SARTHI in Gujarat, have introduced flexible work schedules which allow women to vary their working hours and the amount of work they do each day to fit their busy schedules (Sarin and Khanna, 1990: 21).

Plantation

Women's participation in plantation activities follows the same pattern as their participation in land preparation activities.

Protection

Whereas women have pioneered efforts to protect forests, they are often not assigned to protection work in joint protection or plantation schemes.[46] In some Forest Protection Committees (FPCs) in West Bengal, men are assigned on a voluntary rotation basis to protect the forests. The leader of one such FPC reported that men do all the patrolling although, as he admitted: 'Traditionally, women go into the forest every day while men go only five or six times a year.' When asked why women, who can readily identify offenders in the forest, are not used for patrolling, the leader replied: 'We use them as informants.' There were 70 men and five women members in that Committee: the five women were female-heads of households.

In some of the FPCs, women assume the role of 'unrecognised' protectors, whether or not they are formally assigned to do so. This is because they visit the forests each day. As observed by one woman, 'there is no time in the day when a few women are not present in the forest' (Viegas and Menon, 1990: 26). In response, many FPCs, although they do not officially recognise the women as patrollers, consider it unnecessary to appoint special patrols to guard the forests.

However, some projects have recognised women's special commitment to protect forests and plantations. In Dharwar district of Karnataka State, the Forest Department recruited a women's dairy cooperative (organised by India Development Services, a local NGO) to protect sections of a thousand-acre *anjun* plantation.[47] Earlier, the plantation had been poorly protected and managed

[46] In exclusive all-women schemes, women patrol plantations or forests on a paid or voluntary rotation basis. When they are not able to do so, they hire a watchman (Burra, 1990b: 15).

[47] Anjun is a drought-resistant source of fodder (Burra, 1990c: 12)

by the Forest Department, and villagers were indiscriminately cutting leaves and branches. In her Mission Report to the ILO, Neera Burra who visited this project describes what happened as follows:

> The women's dairy cooperative society wrote to the District Forest Officer about the rapid degradation of the plantation. Since the people were cutting off branches as well as the tender leaves, no *anjun* seeds were available for nursery raising in the area. The Forest Department responded immediately and held training programmes where about 20 people came and were shown how the leaves should be cut so that the tree can grow as well. The women's dairy cooperative society has been given 100 acres to manage and the regeneration of this area was remarkable, with mere protection. Last year the villagers were able to collect *anjun* seeds which they sold to the Forest Department. Because of the initiative taken by the women's group, they were given responsibility of a large plantation. The idea is that once this tract regenerates, it will be opened for access and another portion of the plantation will be given to them to protect. In this manner, the 1,000-acre *anjun* plantation will be regenerated. (Burra, 1990c: 12)

In Himachal Pradesh more than 25 *mahila* guards have been recruited by the Forest Department. However, these women are not used to guard plantations or forests. Rather, they are used as motivators to promote involvement of women in social forestry activities (Dioman, 1989: 5).

Maintenance

Maintenance activities are most often carried out by voluntary labour as project funds typically cover only the establishment phase of plantations. As in other tasks when voluntary labour is called for, many projects report that women predominate in maintenance operations (weeding, pruning, and watering). Further, many projects report that women are more interested and take greater responsibility in ensuring that the plantations are successful (Sarin and Sharma, 1990: 18).

Collection and harvesting

Given their traditional roles as primary collectors of NTFP, women often play primary roles in harvesting under wasteland development projects. Different harvesting arrangements have been worked out in different projects. In the Garhwal region of the hills of Uttar Pradesh (U.P.), one group of women who established a mixed plantation decided that they would harvest intermediate products from the plantation twice a year. Each time, each household is allowed to send one woman for seven days to collect grass and for two days to collect fuelwood. This arrangement is strictly enforced. If more than one woman per household comes to harvest grass or fuelwood, their sickles are taken away (Burra, 1990b: 15)

But men sometimes object to women harvesting. An NDDB official describes what happened in one Tree Growers' Cooperative as follows:

> The cooperative decided that the sale of grass would commence from mid-November. Since cutting and baling grasses was working out to be too expensive, the committee decided to allow members to cut grass, have it weighed and take it away after weighing. Our team had suggested that only women be allowed to cut grass, and that one woman Management Committee member, who was literate, make a record of all sales and cash transactions during that time since the Secretary of the Society was away at NDDB for training. The Management Committee took it lightly and did not comply, saying that a woman could not manage such things, and that male committee members would do the job in turns. What followed then was a growing chaos of men and women frantically cutting grass and fighting over the best area. The supervision was weak because except one member, none of the men from the committee turned up for supervision. When things went out of hand the grass cutting was stopped. Our team went to the village subsequently and held a meeting in which the members themselves admitted that they had committed a mistake and that our earlier suggestion of only women cutting grass under the supervision of a woman Committee Member would produce better results (Mishra, 1989: 21).

Since most plantation or protection schemes have not yet matured, the degree to which women will be involved in the final harvest is not yet known. Men are the primary fellers and loggers of timber for the Forest Departments.

Processing

Since most plantations have not yet matured, the production of many products has not reached the point where large-scale processing is required. Many projects plan to set up processing units in the future, including: oil presses, rope-making units, drying-storage units, and silk spinning and reeling units. In the interim, processing is carried out by individual households using traditional technologies and labour inputs. Although there are regional variations, which should be taken into account for planning purposes, women are the traditional processors of most NTFP, including: cleaning, drying, and storing *mahua* flowers; collecting and drying seeds; cutting and bundling *datun* (reeds used for cleaning teeth); collecting and bundling *tendu* leaves; and collecting and stitching *sal* leaves into leaf-plates (Kaur, 1988: 40). And female labourers predominate in many resource-based industries: notably, rolling *beedis*.

Whether they process NTFP on a self-employed or wage labour basis, women's remuneration is very low (SIDA, 1989b). If employed as wage labour by the Forest Departments or in resource-based industries, women are paid on a piece-rate basis: so many rupees for so many bundles of leaves, leaf-plates, or rolled *beedis*. A few projects have unionised women to demand higher wages. DISHA, and NGO in Gujarat, recently won a court battle to raise the piece-rate for bundling *tendu* leaves to the minimum wage in Gujarat by documenting how much time was required to bundle the leaves.

Marketing

Again, since most plantations have not yet matured, most projects have not yet marketed the final produce. Under most projects, individual households market intermediate products through traditional marketing channels. A few projects have tried to negotiate alternative markets and better prices for women (or for local people more generally).

Benefit or profit-sharing

The issue of how (i.e., to whom) benefits are distributed is almost as central as the issue of how (i.e., in whose name) tenurial rights are negotiated. As in the case of tenurial rights, women can receive benefits either individually or as members of groups and either directly through exclusive all-women projects, jointly with men under joint projects, or indirectly through men under joint projects. To illustrate, women could become joint signatories to the contract signed between individual FPCs and the Forest Department in West Bengal: whereby they would become direct, legal, joint beneficiaries to the profits realised under the contract (100 per cent of the value of NTFP and 25 per cent of the timber value). As noted above, women are joint signatories (and, thereby, direct beneficiaries) to a contract negotiated between local communities and the Forest Department of Haryana for the rights to collect *banjara* grass. If women are not included as signatories to legal contracts, they can at best exercise indirect non-enforceable claims on project benefits.

Monitoring and evaluation

Again, women can have either a direct, joint, or indirect voice in participatory monitoring and evaluation.

Other interventions

Two other topics are often raised in any discussion of women and wasteland development. First, the issue of fuel-saving devices: notably, improved wood-burning stoves, biogas units, and solar cookers or heaters. Second, the issue of how women can be involved in forestry extension, as either extension staff or clients. Both are important issues which need to be promoted and addressed. But neither of these issues should serve to divert attention from the more central issue of involving women in mainstream wasteland development projects. Project experience suggests that female extension agents could help ensure that project information reaches women and that women participate more actively in projects (Olsson, 1988). However, project experience also indicates that women's participation can be increased without the presence of female extension agents.

What Happens when Women are Involved

There are three types of possible impact from involving women in wasteland development: on the project itself, on the women themselves, and on the welfare of their children and households. As noted above, many projects report that women's responsiveness and sense of responsibility have served to increase the productivity and viability of wasteland development projects (Sarin and Sharma, 1990). Conversely, when women are not fully involved, projects are not as successful as hoped.

Many projects report that involving women has served not only to meet some of women's practical needs (e.g., reduced time and effort to collect fuel and fodder) but also some of women's strategic needs (e.g., tenurial rights and political representation) and some of women's empowerment needs (e.g., self-confidence, respect, and power). In their case study of SARTHI in Gujarat, Sarin and Khanna report that getting organised as a group and undertaking collective action clearly had an empowering effect on the women participants (Sarin and Khanna, 1990: 22). This empowerment had several dimensions: the women perceived and established their rights to information, to legitimised roles, including leadership roles, in the public domain, and to respect and support within their families or villages (ibid.).

Finally, many projects report that involving women has served to increase the welfare of the women's households. This is largely because women typically manage the daily subsistence and nutritional needs of households and typically pool more of their income (than men do) into the household budget. As one analyst, in summarising diverse project experience, noted: 'if forest produce goes primarily to men's (cash-generating) enterprises and exacerbates shortages of products required by women for either subsistence needs or market-oriented activities, the net income accruing to the household may decline rather than increase' (Molnar and Schreiber, 1989: 12, Box 8).

Section IV: Restoring the Balance

This paper has analysed effort within India to restore the balance not only between people and natural resources through wasteland development but also, in so doing, between men and women. This concluding section presents a conceptual framework for analysing women's participation in wasteland development and argues that three strategic needs of women need to be addressed, namely: women's tenurial rights, women s representation in local institutions, and gender-based planning.

Women and Wasteland Development: A Conceptual Framework

The paper has set out three factors which are critical to the full and effective involvement of women in wasteland development: participation of women in all stages of wasteland development; use or ownership rights to women over land to be developed or over products from that land; and representation of women in local institutions established to manage local resources. Women's participation, women's tenurial rights, and women's representation can be negotiated in either of three basic ways: **directly** (in their own right); **jointly** (with men from their households or villages); or **indirectly** (through men from their households or villages). In general, women benefit least from indirect participation; benefit only partially through joint participation; and benefit fully when they are recognised as legitimate clients, right-holders, and representatives in their own right.

As a tool for designing, implementing, and evaluating wasteland development projects which seek to involve women, this framework is presented in the following matrix. In this matrix, the key variables in involving women are paired with the different modalities of involving women. In the horizontal axis, women's participation is differentiated by whether women are seen as labourers, as legitimate clients, or as full partners (owner-managers) in wasteland development and by whether their participation is direct, joint, or indirect (as described above). In the vertical axis, three broad stages of participation are listed: design, implementation, and management. A more complete listing would include, of course, all the sub-stages described in Section IV on Women and Wasteland Development. Also, in the vertical axis, three types of rights are listed: ownership rights to land, usufruct rights to land, and beneficiary rights to products from the land. And three types of local institutions are listed: all-women institutions, joint (male and female) institutions, and separate (male or female) institutions.

Framework for Analysing Women's Participation

	Labourers	*Clients*	*Partners*	*Direct*	*Joint*	*Indirect*
Participation: design implementation management						
Rights: ownership usufruct benefits						
Institutions: all-women joint separate						

Strategic Gender Needs

In a recent paper, Moser draws a useful distinction between practical gender needs and strategic gender needs. She defines practical gender needs as: 'those needs which are formulated from the concrete conditions women experience, in their engendered position within the sexual division of labour, and deriving out of this their practical gender interests for human survival' (Moser, 1989: 1803). And she defines strategic gender needs as: 'those needs which are formulated from the analysis of women's subordination to man, and deriving out of this the strategic gender interest identified for an alternative, more equal and satisfactory organisation of society than that which exists at present, in terms of both the structure and nature of relationships between men and women' (ibid.). As Moser argues, strategic and practical gender needs will vary depending on the particular cultural and socio-political context within which they are formulated. Typically, strategic gender needs include the removal of institutionalised forms of discrimination, such as the right to own land or property and political equality (ibid.).

In the context of wasteland development in India, practical gender needs include women's need for secure sources of fuel, fodder, and water and women's need for an independent source of income. And strategic gender needs include, primarily, women's need for secure rights over natural resources and direct representation in local resource-management institutions and, by extension, the need for gender-based wasteland development and land use planning. Because they generally involve new land use and institutional arrangements, wasteland development projects offer a unique opportunity to promote these basic strategic needs of women.

In countries, such as India, where women's rights to private land are limited, women's rights to common land become critical. Because few women have rights to private land, biomass goods from common lands are often the only source of independent income for women. However, because traditional use rights to common lands are vested in households or male-heads of households, women exercise tenuous claims to biomass goods from common lands. When new tenurial rights are being negotiated, as in wasteland development, women's rights to common land should be legalised. Establishing women's rights to common land could serve to legitimise the concept of women's rights to property and could represent, therefore, a first step in negotiating women's rights to private land (Saxena, 1987: 19).

Wasteland development efforts often involve rebuilding village *panchayats* or building alternative local institutions (Singh, 1986: 9). During this process of democratisation, there are opportunities to promote women's political representation and voice. Wasteland development projects can effectively promote women's participation in local institutions (and the public domain more generally) by offering women the opportunity to move outside their villages, to interact with other women, to receive training in various skills, to attend meetings on a regular basis, and to develop the self-confidence to speak out at meetings and assume leadership roles (Sarin and Khanna, 1990: 31–32). Promoting both women's tenurial and political rights can be mutually reinforcing, as greater decision-making and leadership roles often flow out of owning or controlling assets (Krishnan, 1989: 18).

This paper presents a framework for gender-based wasteland development planning. The key variables of this framework are summarised in the matrix above. In the context of wasteland development, gender-based planning should serve to promote the primary strategic needs of women: rights to the land being developed or to products from that land; and political representation in local institutions established to manage that land. The promotion of women's tenurial rights and women's political representation through gender-based wasteland development can be seen as an end in itself. It can also be seen as a means to promoting strategic gender needs in other sectors, including: women's rights to private agricultural land; women's political representation in other local institutions, notably village *panchayat*; and gender-based land use planning or development planning more generally. In designing its programme on women and wasteland development, the ILO explicitly adopted the strategic objective of empowering women by securing women's rights to land and by promoting local women's organisations.

APPENDIX 1.1: LIST OF PROJECTS REVIEWED

Projects Visited by Neera Burra (ILO) and Martha Chen (HIID)

PEDO People's Education and Development Organisation (Rajasthan)
Seva Mandir (Rajasthan)
Ubeshwar Vikas Mandal (Rajasthan)
Centre for Women's Development Studies (West Bengal)
Forest Protection Committees (West Bengal)
Gram Vikas (Orissa)
PREM People's Rural Education Movement (Orissa)

Additional Projects Visited by Martha Chen

Aga Khan Rural Support Program (Gujarat)
SEWA Self-Employed Women's Association (Gujarat)
Mahiti (Gujarat)

Additional Projects Visited By Neera Burra

Bhagavatula Charitable Trust (Andhra Pradesh)
Central Himalayan Environment Association (Uttar Pradesh)
Jakeshwar Sikshan Sansthan (Uttar Pradesh)
Dasholi Gram Swarajya Mandal (Uttar Pradesh)
Rural Integrated Development Organisation (Tamil Nadu)
Association for Rural Community Development (Tamil Nadu)
India Development Services (Karnataka)

Case Study Projects

PEDO People's Education and Development Organisation (Rajasthan)
SARTHI Social Action with Rural and Tribal Inhabitants of India (Gujarat)
Forest Protection Committees (West Bengal)
Jakeshwar Sikashan Sansthan (Uttar Pradesh)

Legal Case Studies

Himachal Pradesh
Gujarat

REFERENCES

Aga Khan Rural Support Programme (India) (1989) *Proceedings of the Workshop on 'Common Property Land Resources (CPR-Land) in Gujarat and problems of their development'.* Organised by the Aga Khan Rural Support Programme (India), the Gujarat Institute of Area Planning, Ahmedabad and the Revenue Department of the Government of Gujarat, 17 March.

Agarwal, Anil (1982) 'Firewood: Fuel of the rich?' *Earthscan Bulletin,* Vol. 5, No. 4, 5 July.

——— (1984) *'Beyond pretty trees and tigers: The role of ecological destruction in the emerging patterns of poverty and people's protests.'* Fifth Vikram Sarabhai Memorial Lecture delivered at New Delhi, Indian Council of Social Science Research, 1 August.

——— (1985a) 'The environment and rural women.' *Indian Express Magazine,* 2 June.

——— (1985b) 'Population and environment.' *Economic and Political Weekly,* Vol. XX, No. 24, 15 June.

——— (n.d.) *Strategies for the involvement of the landless and women in afforestation: Five case studies from India.* New Delhi: Centre for Science and Environment.

——— (1987) 'Between need and greed: The wasting of India, the greening of India.' In Anil Agarwal, D. D'Monte and U. Samarth (Eds.), *The fight for survival.* New Delhi: Centre for Science and Environment.

Agarwal, Anil and B. Bhatt (1983) *Firewood in cities: I and II.* Reports No. 112 and 113. New Delhi: Centre for Science and Environment.

Agarwal, Anil and P. Deshingkar (1983) *Headloaders: Burdens of energy—II.* Report No. 119. New Delhi: Centre for Science and Enivornment.

Agarwal, Anil and Sunita Narain (1985a) *Strategies for the involvement of the landless and women in afforestation*. New Delhi: Centre for Science and Environment.

——— (1985b) 'Women and natural resources.' *Social Action*, Vol. 35, No. 4, October–December.

——— (1989) *Towards green villages: A strategy for environmentally-sound and participatory rural development.* New Delhi: Centre for Science and Environment. September.

Agarwal, Bina (1986a) *Cold hearths and barren slopes: The woodfuel crisis in the Third World.* Riverdale, Maryland: Riverdale Publishers and Delhi: Allied Publishers.

——— (1986b) *Under the cooking pot: The political economy of the domestic fuel crisis in rural South Asia.* Delhi: Institute of Economic Growth.

Arnold, J.E.M. and W.C. Stewart (1989) *Common property resource management in India.* Report to the India Agriculture Division. Washington, D.C.: World Bank. June.

Arnold, Mike, A. Bergman, P. Harris, and J. Mohanty (1989) *'Evaluation of the SIDA supported social forestry in Orissa, India.'* Background Paper for Workshop on Involvement of Women in Social Forestry. New Delhi: NWDB.

ASTRA (1981) *Rural energy consumption pattern—A field study*. Bangalore: Indian Institute of Science.

Bandyopadhyay, J. and Vandana Shiva (1986) 'Drought, development and desertification.' *Economic and Political Weekly,* 16 August.

Batliwala, S. (1983a) 'Women and cooking energy.' *Economic and Political Weekly*, Vol. XVIII, No. 52–53.

——— (1983b) 'Rural energy scarcity and under-nutrition: A new perspective.' *Economic and Political Weekly*, Vol. XVIII, No. 9.

Bhaduri, T. and V. Surin (1980) 'Women "headloaders" of Ranchi.' *Voluntary Action*, Vol. 23, No. 5.

——— (1980) 'Community forestry and women headloaders.' In *Community Forestry and People's Participation*, Seminar Report, Ranchi Consortium for Community Forestry. November.

Bhati, J.P. and D.V. Singh (1987) 'Women's contribution to agricultural economy in hill regions of north-west India.' *Economic and Political Weekly*, Vol. XXII, No. 17, 25 April.

Bhumbla, D.R. (n.d.) *Fuelwood plantations*. New Delhi: Society for Promotion of Wastelands Development.

Bhumbla, D.R. and Arvind Khare. (n.d.) *Estimate of wastelands in India*. New Delhi: Society for Promotion of Wastelands Development.

Blair, Harry W. (1986) 'Social forestry: Time to modify goals?' *Economic and Political Weekly*, Vol. XXI, No. 30, 26 July.

Brara, Rita (1987) *Shifting sands: A study of rights in common pastures*. Jaipur: Institute of Development Studies.

Brokensha, David (1988) '*Village-level management of common property resources, especially fieldwood and fodder resources, in Karnataka, India*.' Report prepared for the World Bank.

Bromley, D.W. and **N.M. Cernea** (1989) *The management of common property natural resources: Some conceptual and operational fallacies*. World Bank Discussion Papers No. 57. Washington, D.C.: The World Bank.

Burra, Neera (1990a) *Mission Report: 20 February–8 March 1990*. New Delhi: ILO.

——— (1990b) *Mission Report: 22 May–1 June 1990*. New Delhi: ILO.

——— (1990c) '*Mission Report: 7 June–4 July 1990*.' New Delhi: ILO.

——— (1990d) *Women and wasteland development: The ILO perspective*. New Delhi: ILO.

Caldwell, John, P.H. Reddy and P. Caldwell (1988) *The causes of demographic change-experimental research in south India*. Madison: University of Wisconsin Press.

Carr-Harris, Jill (1990) '*Jakeshwar Sikshan Sansthan: Can it meet the challenge? Changes in the hill economy, forest cover and status of women in Dasholi block, Garhwal*'. A case study prepared for the ILO Technical Workshop on Women and Wasteland Development. New Delhi: ILO.

Ceceleski, Elizabeth (1985a) *The rural energy crisis, women's work and family welfare: Perspectives and approaches to action. (World employment research)*. Rural Employment Policy Research Programme. Geneva: ILO.

——— (1985b) *The rural energy crisis: Women's work and basic needs*. Rome: FAO.

Centre for Science and Environment (CSE) (1982) *The state of India's environment [1982]: A citizen' report*. New Delhi: Centre for Science and Environment.

——— (1985) *The state of India's environment [1984-85]: The second citizen's report*. New Delhi: Centre for Science and Environment.

——— (1987) *The wrath of nature, the impact of environmental destruction on floods and droughts*. New Delhi: Centre for Science and Environment.

Cernea, Michael M. (1989) *User groups as producers in participatory afforestation strategies*. World Bank Discussion Paper No. 70. Washington, D.C.: The World Bank.

Chambers, Robert, N.C. Saxena and T. Shah (1989) *To the hands of the poor: Water and trees*. New Delhi: Oxford and IBH Publishing Co.

Chand, Malini and Rekha Bezboruah (1980) 'Employment opportunities for women in forestry.' In H.N. Mathur et al. (Eds.) *Role of women in community forestry*. Dehradun.

Chandra, Ramesh (1989) *Perspective of social security plantations*. Background Paper for Workshop on Involvement of Women in Social Forestry. New Delhi: National Wasteland Development Board.

Chavangi, Noel A. (1988) *Case study of women's participation in forestry activities in Kenya*. Prepared for Community Forestry Office, Forestry Department. Rome: FAO.

Chen, Martha (1991) *Coping with seasonality and drought*. New Delhi: Sage.

Chowdhry, Kamla (1979) *Non-formal education and development*. Presented at a symposium on 'Agricultural research and education systems for development', Golden Jubilee Celebrations of the Indian Council of Agricultural Research, 5 September.

——— (1988) *Poverty, environment, development: People's participation and people centred structures, glimpses from Gandhian thinking*. May.

CIDA (1988) *Andhra Pradesh social forestry project: Mid-term review*. March. New Delhi: Canadian International Development Agency.

Damodaran, A. (1987) 'Structural dimensions of fodder crisis: A village study in Karnataka.' *Economic and Political Weekly*, Vol. XXII, No. 13, Review of Agriculture, 28 March

Das, S.K. (1989) 'Welfare of rural women through social forestry—West Bengal experience.' Paper presented at Workshop on 'Involvement of Women in Social Forestry'. New Delhi: National Wasteland Development Board.

Dasgupta, Subhachari (1986) 'Forest, ecology and the oppressed.' New Delhi: People's Institute for Development and Training (PIDT).

Dioman, D.R. (1989) 'Involvement of women in social forestry in Himachal Pradesh.' Paper presented at Workshop on 'Involvement of Women in Social Forestry'. New Delhi: National Wasteland Development Board.

Dover, Michael and Lee M. Talbot (1987) *To feed the earth: Agro-ecology for sustainable development*. New York: World Resources Institute.

Fernandes, W. (1988) *Deforestation, forest dweller economy and women*. New Delhi: Indian Social Institute.

Fernandes, W. and S. Kulkarni (1983) *Towards a new forest policy: People's rights and environmental needs*. New Delhi: Indian Social Institute.

Fernandes, W. and G. Menon (1987) *Tribal women and forest economy*. New Delhi: Indian Social Institute.

Fernandes, W. et al. (1988) *Forests, environment and tribal economy: Deforestation, impoverishment and marginalisation in Orissa*. New Delhi: Indian Social Institute.

Food and Agriculture Organisation of the UN (FAO) (1989) *Women in community forestry (a field guide for project design and implementation)*. Rome: FAO.

FAO and SIDA (1989) *Community forestry: Rapaid appraisal.* Community Forestry Note 3. Rome: FAO.

——— (1987) *Restoring the balance: Women and forest resources*. Rome: FAO.

Ford Foundation (1989) *Joint management of forest lands: Experiences from South Asia*. New Delhi: Ford Foundation.

Fortmann, Louise (1986) 'Women in subsistence forestry.' *Journal of Forestry*, Vol. 84, No. 7.

Fortmann, Louise and J. Riddell (1985) *Trees and tenure: An annotated bibliography for agroforesters and others*. Madison, Wisconsin: Land Tenure Centre, University of Wisconsin; and Nairobi, Kenya: ICRAF. January.

Fortmann, Louise and D. Rocheleau (1985) *Women and agroforestry: Four myths and three case studies*. Reprint No. 19, Nairobi, Kenya: ICRAF.

Fortmann, Louise and John Bruce (Eds.) (1988) *Whose trees?* Boulder, Colorado: Westview Press.

Foundation to Aid Industrial Recovery (1984) *An urban fuelwood plantation for Delhi: A Proposal*. May.

Gadgil, Madhav (n.d.) *Community based resource management systems*. Background Note for the Meeting of the Steering Group on Forests, Environment, and Wasteland Development.

GOI (1973) *Interim report of the National Commission on Agriculture and Social Forestry*. Ministry of Agriculture, Government of India.

——— (1988a) *Shramshakti*. Report of the National Commission on Self-Employed Women and Women in the Informal Sector.

——— (1988b) *National Forest Policy, 1988*. Ministry of Environment and Forests, Department of Environment, Forests and Wildlife, 7 December.

Gupta, Tirath and Amar Guleria (1982) *Some economic and management aspects of a non-wood forest product in India: Tendu leaves*. New Delhi: Oxford and IBH.

Hoskins, Marilyn (1979) *Women in forestry for local community development: A programming guide*. Washington, D.C.: Office of WID, United States Agency for International Development. September.

——— (1983) *Rural women, forest outputs and forest products*. Rome: FAO.

ILO (1987) *Wasteland development through women's organisations*. Project document. New Delhi: ILO.

——— (1988a) *Employment and income generation through social forestry in India: Review of issues and evidence*. New Delhi: ARTEP.

——— (1988b) *The Bankura story: Rural women organise for change*. New Delhi: International Labour Organisation.

ISO/SWEDFOREST (1987) *Workshop on role of incentives in social forestry*. Background Papers. New Delhi.

Jain, Shobhita (1984) 'Women and people's ecological movement: A case study of women's role in the Chipko movement in Uttar Pradesh'. *Economic and Political Weekly*, Vol. XIX, No. 41, 13 October.

Jha, Prem Shankar (1983) 'Farm forestry under attack—ecological argument unfounded.' *The Times of India*, 1 August.

Jodha, N.S. (1983) *Market forces and erosion of common property resources*. Hyderabad; ICRISAT.

——— (1985) 'Population growth and decline of common property resources in Rajasthan, India.' *Population and Development Review* , Vol. II. No. 2, June, ICRISAT Journal Article No. 441.

——— (1986) 'Common property resources and rural poor in dry regions of India'. *Economic and Political Weekly*, Vol. XXII, No. 27.

Kamweti, David (1982) *Tree planting in Africa, south of the Sahara*. Nairobi, Kenya: The Environment Liaison Centre.

Kaur, Ravinder (1988) *Women's role in forestry in India*. Consultant Report for the India Country Women in Development Study. Washington, D.C.: World Bank, PHRWD. September.

Kearns, Virginia (1985) *Monitoring of women's role in forestry projects*. Rome: FAO.

Khetan, Neelima (1989) *Wasteland development: A review (1986–89)*. Udaipur: Seva Mandir.

Kishwar, Madhu (1987) 'Toiling without rights: Ho women of Singhbhum'. *Economic and Political Weekly*, Vol. XXII, No. 3, 17 January.

Krishnamurthy, K.G. (1989) *Involvement of women in social forestry*. Background Paper for Workshop on Involvement of Women in Social Forestry. New Delhi: National Wasteland Development Board.

Krishnan, V.S. (1989) *NWDB schemes: Institutional support and operational strategies*. Background Paper for Workshop on Involvement of Women in Social Forestry. New Delhi: NWDB.

Kumar, Shubh K. and David Hotchkiss (1988) *Consequences of deforestation for women's time allocation, agricultural production, and nutrition in hill areas of Nepal*. Research Report No. 69, Washington, D.C.: IFPRI. October.

Kumar, Nalini (1988) *The role of women in forestry and natural resource management*. Consultant Report. Washington, D.C.: World Bank, PHRWD. August.

Leach, G. and M. Gowen (1966) *Household energy manual: An interim guide and reference manual*. Washington, D.C.: World Bank.

Lekha, Chandra ((1989) *Tree patta scheme—An experiment in Tamil Nadu*. Background Paper for Workshop on Involvement of Women in Social Forestry. New Delhi: NWDB.

Maiti A.K. and S. Dasgupta (1986) *The rural energy crisis: Poverty and women's roles in five Indian villages*. Geneva; International Labour Organisation.

Malhotra, K.C. and Mark Poffenberger (Eds.) (1989) *Forest regeneration through community protection: The West Bengal experience*. Proceedings of the Working Group Meeting on Forest Protection Committees, 21-22 June. Calcutta: West Bengal Forest Department.

Malmer, Pernilla (1987) *Socio-economic change in social forestry: A case study in Kovilur village, Tamil Nadu, India*. Background Paper for Workshop on Involvement of Women in Social Forestry. New Delhi: National Wasteland Development Board.

Mishra, V.K. (1989) *NWDB's experience with the role and involvement of women in their tree grower's cooperatives*. Background Paper for Workshop on

Involvement of Women in Social Forestry. New Delhi: National Wasteland Development Board.

Molnar, Augusta (1986a) *Social forestry experience in India and Nepal: A review of community woodlots, the involvement of women and the introduction of wood-saving devices in the various projects*. Consultant Report. Washington, D.C.: World Bank, ASPAB.

——— (1986b) *Annotated bibliography on women and environment*. Washington, D.C.: World Bank.

——— (1987) *Forest conservation in Nepal: Encouraging women's participation* in *Seeds,* No. 10, New York.

Molnar, Augusta and Gotz Schreiber (1989) *Women and forestry: Operational issues*. Working Papers Series No. 184. Washington, D.C.: World Bank, PHRD.

Moser, Caroline O.N. (1989) 'Gender planning in the Third World: Meeting practical and strategic gender needs.' *World Development*, Vol. 17, No. 11, pp.1799–825.

NABARD (1989) *Role of institutional finance in encouraging women in social forestry*. Background Paper for Workshop on Involvement of Women in Social Forestry. New Delhi: NWDB.

Nath, N.C.B. (1985) *Planning for wastelands development: Some alternative models*. Presented at a Workshop on Wastelands Development in Andhra Pradesh, Hyderabad, 30 August.

National Wasteland Development Board (NWDB) (n.d.) *Wastelands: Challenge and response*. National Wasteland Development Board, Ministry of Environment and Forests, Government of India.

——— (1988) *Agenda papers for the seventh meeting of the NWDB*, Dated 9 May. New Delhi: National Wasteland Development Board.

Ninan, Sevanti (1981) 'Women in community forestry', in the Report of the Seminar on Role of Women in Community Forestry, Ministry of Agriculture and Cooperation: Government of India, Dehradun (mimeo).

Olsson, Gunilla (1988) *Employment and income generation in social forestry: A case study from Orissa*. Geneva: International Labour Office.

Pant, M.M. (1980) *Employment of women in forestry*. Paper presented at the Seminar on Role of Women in Community Forestry. Dehradun: Ministry of Agriculture and Cooperation, Government of India.

Patel V.J. (1985) 'Rational approach towards fuelwood crisis in rural India.' *Economic and Political Weekly*, Vol. XX, No. 32.

FAIR (1983) *Some case studies in social forestry*. New Delhi: Foundation to Aid Industrial Recovery.

Repetto, Robert (1987) 'Population, resources, environment: An uncertain future.' *Population Bulletin*, Vol. 42, No. 2, July.

Rocheleau, Dianne E. (1988) 'Gender, resource management and the rural landscape: Implications for agroforestry and farming systems research' in Susan Poats et al., (Eds.), *Gender issues in farming systems research and extension*. Boulder, Colorado: Westview Press.

Saint, Kishore (1989) 'Aravalli Bhils show the way.' *Indian Express*, 29 September.

Sarin, Madhu and Renu Khanna (1990) *Wasteland development by women's groups: A case study of SARTHI in Santramputr taluka of Panchmahals district, Gujarat.*

A case study prepared for the ILO Workshop on Women in Wasteland Development. New Delhi: ILO.

Sarin, Madhu and Chandrika Sharma (1990) *Wasteland development by women's groups: A case study of PEDO in Bicchiwara block of Dungarpur District, Rajasthan*. A case study prepared for the ILO Workshop on Women in Wasteland Development. New Delhi: ILO.

Saxena, N.C. (1987) *Participatory planning for wastelands development*. New Delhi: Council for Social Development.

Scott, Gloria (1980) *Forestry projects and women: Sectoral overview*. Washington, D.C.: The World Bank.

——— (1982) 'Forestry projects: How women can help—and help themselves.' *Appropriate Technology*, Vol. 9, No. 3.

Shah, Parmesh (1987) *Potential of agroforestry as land use technology for rural poor*. New Delhi: Aga Khan Foundation.

Sharma, Kamal (Ed.) (1984) *Economics of wastelands development*. Proceedings of SPWD–NABARD Seminar, Suraj Kund, India, March.

Shingi, P.M., M.S. Patel and S. Wadhwalkar (1986) *Development of social forestry in India*. New Delhi: Oxford and IBH Publishing Co.

Shiva, Vandana (1983) 'People and forests: Crucial fight ahead.' *Economic and Political Weekly*, Vol. XVIII, No. 13.

——— (1986) 'Coming tragedy of the commons'. *Economic and Political Weekly*, Vol. XX, No. 15, 12 April.

——— (1988) *Staying Alive: Women, ecology and survival in India*. New Delhi: Kali for Women.

SIDA (1989a) *Participation of women in social forestry*. Background Paper for Workshop on Involvement of Women in Social Forestry. New Delhi: NWDB.

——— (1989b) *Employment and income generation through social forestry in India: Review of issues and evidence*. Background Paper for Workshop on Involvement of Women in Social Forestry. New Delhi: NWDB.

Singh, Chhatrapati (1986) *Common property and common poverty: India's forests, forest dwellers and the law*. Delhi: Oxford University Press.

——— (1990) *Women's involvement in wasteland development and the law*. Legal Study prepared for the ILO Workshop on Women and Wasteland Development. New Delhi: ILO.

Singh, Chhatrapati, T. Paul and P. Prinja (1986) *Wasteland development and the law*. Bombay: N.M. Tripathi Pvt. Ltd.

Singh, R.P. and M. Osman (1987) *Agroforestry systems for small holdings*, International Union of Forestry Research Organisation, Workshop on Agroforestry for Rural Needs, New Delhi, 22–26 February.

Singh, R.V. (1990) *Fodder trees of India*. New Delhi: Oxford and IBH Publishing Co.

Skutch, Margaret (1987) *Why people don't plant more trees*. Washington, D.C., World Resources Institute.

Society for Participatory Research in Asia (1984) *Deforestation in Himachal Pradesh*, New Delhi, Society for Participatory Research in Asia, Khanpur, June.

Society for Promotion of Wastelands Development (1986) *National workshop on landless people and wastelands development*, New Delhi, April.

Spears, John and Scott Guggenheim (1986) *Sociological dimension of social forestry projects*. Washington, D.C.: World Bank, ENVST, 31 August.

Tewari, R.N. (1981) *Effective policies and legislative needs for stimulating growth through community forestry in forest-based economies within an agrarian environment—The case of India*. Invited Paper to XVII World International Union of Forestry Research Organisation, Japan.

Tinker, Irene (1980) *Women and energy: Program implications*. Office of Women in Development. Washington, D.C.: USAID.

Verma, D.P.S. (1988) 'Women and forest nurseries in Ganeshpura village.' *The International Tree Crops Journal*, Vol. 5. Great Britain, A.B. Academic Publishers.

——— (1989) *Individual beneficiary scheme of the Gujarat Forest Department*. Background Paper for Workshop on Involvement of Women in Social Forestry. New Delhi: NWDB.

Viegas, Philip and Geeta Menon (1990) *Forest protection committees of West Bengal: Role and participation of women*. A case study prepared for the ILO Workshop on Women and Wasteland Development. New Delhi: ILO.

Vohra, B.B. (1988) *Managing India's land resources: A vital but neglected task*. Text of the Nehru Memorial Lecture. New Delhi, India, Teen Murti House. 26 May.

Wiersum, K.F. (1986) *Social forestry and agroforestry in India*. Rijksinstituut voor onderzoek in de Bos- en Landschapsbouw 'De Dorschkamp', Wageningen, the Netherlands, Rapport nr. 456.

World Resources Institute (1985) *Tropical forests: A call to action*. Report prepared by International Task-Force convened by World Resources Institute, the World Bank and the United Nations Development Programme: Washington D.C.

2 *Experiments in the Field: The Case of PEDO in Rajasthan*

MADHU SARIN
CHANDRIKA SHARMA

Introduction

This is a case study of women's involvement in wasteland development. It is based on the work of the People's Education and Development Organisation (PEDO) in Bicchiwara block of Dungarpur district, Rajasthan.

PEDO started working in Bicchiwara block in June 1980, as a branch of the Social Work and Research Centre (SWRC), Tilonia. It obtained independent registration as PEDO in 1986. During the last ten years, the organisation has initiated a wide range of development activities in the block. Some of the major ones have been: eradication of guinea worm, installation of hand pumps for safe drinking water, digging and deepening irrigation wells including installation of community wells and lift irrigation systems, running schools for children, contour bunding of agricultural fields, a soil conservation programme and a major environmental programme for the rehabilitation of degraded private and community wastelands.

Starting from giving training to women in handloom and *durree* production, PEDO's Women's Programme has developed into a major programme over the years with a shift in emphasis from project-based activities to organising and mobilising women to deal with their problems. The first effort in this direction was to mobilise women to plant trees on private wastelands as a long-term measure to counter the disastrous impact of drought. Hostile male reactions to the PEDO's emphasis on working only with women on a land-based activity led to a change in approach. Combined groups of men and women were formed to develop degraded village common lands. Despite women's greater stake in augmenting fuel and fodder resources, their participation in the combined groups remained limited till independent women's groups were formed. These have empowered women to take greater initiative in village affairs. This now includes their dynamic participation in the management of common lands in many villages.[1]

The Area

Dungarpur district falls in the semi-arid belt of southern Rajasthan. Out of its total population of 682,845 (Census of India, 1981), 440,026 (64.4 per cent) are Bhil tribals (referred to as *adivasis* hereafter) and 30,828 (4.4 per cent) belong to Scheduled Castes. The average annual rainfall in the district is 762 millimetres, most of which falls during the monsoon months of July and August. The yearly variation in rainfall is large due to which the district is known to be drought prone. Historical records indicate the occurrence of several major droughts in this century. The most recent drought from 1985 to 1987 was the worst since 1891.

The terrain consists of low, rolling hills. In the local Wagri language, *dungar* means a small hill from which the district probably derives its name.

Till Independence, the district was heavily forested. Some were primarily teak forests and others mixed deciduous forests of *teak, tendu, dhawra, bamboo* and a few other species.

[1] As members of a small professional resource group, the co-authors of this case study have been associated with PEDO since 1983 and 1986 respectively. We are thus writing this case study not as outsiders, but as participants in a rich learning process spanning several years.

We would like to express our gratitude to Devilal Vyas (Director, PEDO), Ramilla Vyas (In-charge, Women's Programmes), Balraj Singh (In-charge, Environmental Programme), all the women staff, field staff and a large number of village men and women for all their assistance and cooperation.

During the four and a half decades since Independence, the district has witnessed devastating deforestation combined with significant changes in land use patterns. The following statistics give an indication of the change:

Table 2.1
Changes in Land Use in Durgapur District over Four Decades

	1943–44	*1985–86*
Total area	379,200 ha	379,200 ha
Population	286,000	682,000
Cultivated area	41,300 ha	130,097 ha
Forest area	332,348 ha	61,815 ha (area under dense forest only about 5,000 ha)
Uncultivated wasteland		77,614 ha
Pasture land		39,886 ha

Source: *Wasteland News,* Vol. II, No.1,August–October1986. p.6.

More recent statistics are likely to be even more alarming as deforestation was further accelerated during the prolonged drought from 1985 to 1987.

Bicchiwara Block

Bicchiwara is one of five administrative blocks in the district. Out of the block's total population of 138,264 (Census of India, 1981), 81.5 per cent are *adivasis* (Bhils) and 2.3 per cent belong to Scheduled Castes. The majority of the *adivasis* practise rainfed agriculture on unterraced hill slopes.

Compared to an all-India literacy rate of 36 per cent, the literacy rate in Dungarpur district is only 18.52 per cent and that in Bicchiwara block only 13.7 per cent. The literacy rate among women in the block is a mere 3.84 per cent compared to 23.9 per cent for men.

The 1981 Census put the forest area of Bicchiwara block at 5,436 hectares (only 7.6 per cent of the total area). It is estimated that at Independence, over 35 per cent of the block was under dense forest cover.

Dependence of the Local Economy on the Natural Resource Base

The predominantly *adivasi* population of Bicchiwara block historically subsisted by a combination of cultivation and food gathering from the forests. Even today, over 80 per cent of the local population depends on agriculture, most of it rainfed. Agricultural productivity was intimately dependent on the health of the forests which ensured a healthy moisture regime with good replenishment of sub-soil water. This kept wells recharged even during dry periods and there were several perennial springs and streams. Leaf litter from forested slopes fertilised the agricultural fields below. People report that agricultural yields were several times more than those obtainable today. There was abundant grass and leaf fodder to sustain good quality livestock. Till a couple of decades ago, practically every tribal family had one or more buffaloes. These have now been replaced by the hardier goats and low-grade cows.

Besides, forests provided the people with a wide range of subsistence goods—firewood, timber for housing, furniture and agricultural implements, medicinal herbs and plants, edible fruits, gums, honey, roots and wild flowers. During earlier droughts, although there was undoubted hardship, the forests provided a backup life support system. There was an intimately harmonious relationship between the local people and their natural environment. In fact, the Bhils, the oldest inhabitants of southern Rajasthan, have been referred to as *vanputras* or sons of the forest because of their traditional relationship with forests.

Unfortunately, this thickly forested area has today been transformed into a moonscape of bare, brown hills with barely a blade of grass on them. Semi-starved, low-grade cattle, with a preponderance of goats and sheep, can be seen searching for food for survival. With the symbiotic relationship between the forests and the local subsistence economy destroyed, a vicious poverty trap has been created. The bared hill slopes brought under the plough loose rich top-soil each year. The fields in the valleys are deprived of the rich leaf litter coming down the hills leading to reduced yields. If the rains do not come on time, there are few forests to fall back upon for survival. Water sources dry up due to limited recharging of sub-soil water. With less firewood available, precious dung has to be burnt as fuel instead of fertilising the fields. If the crop fails due to inadequate or untimely rainfall, people either have to migrate to other areas in search of work or depend on daily wage work generated through government drought relief programmes. The remaining forests are hacked down further to collect firewood for sale to supplement declining incomes. The traditional subsistence economy is on the verge of collapse with mere survival having become a full-time task.

The Status of Women and their Relationship to the Natural Environment

On the surface, it appears that Bhil women enjoy more freedom than the women of mainstream Hindu society. This stereotype is reinforced by the evident free mobility enjoyed by *adivasi* women and the relative ease with which they can break or establish marital or extra-marital relationships.

There is also a tradition of bride price instead of dowry. This probably emanates from the important contribution of women's labour to the subsistence economy. Besides traditional women's tasks of childbearing, child care and domestic work of cooking, washing and cleaning, there is a clear division of labour between men and women in the subsistence economy. Fetching firewood and water are considered exclusively women's work. Only rarely, if ever, can one see a man fetching either of these. Similarly, taking cattle for grazing is by and large done by women, although children and older men also do this. As there is hardly any tradition of stall-feeding the cattle, fetching fodder is not a major activity. And then, women participate in most agricultural activities such as sowing, hoeing, weeding, harvesting and post-harvest storage and processing. The only agricultural task exclusively done by men is ploughing. When the forests were in good shape, collecting non-wood forest produce other than firewood was also done by women. Today, with people's increasing dependence on wage labour, more women than men can be seen at work sites. Thus, women are also major earners for the family.

One would think that in such a situation, the status of tribal women would be far superior to that of mainstream Hindu women. In reality, as we discovered through years of working with local women, it is little different, if not worse.

Bhil society is strongly partriarchal with a strong preference for the male child. Ownership of all immovable property such as land and housing is vested in men and the only property considered to be women's is their jewellery. There is strong cultural resistance to giving even a part of the family's land to women and a family without male offspring considers it a great misfortune to have to transfer its land to the daughters. In this respect, the Patels, a caste Hindu group, seem more willing than *adivasis* to give a share of the family land to daughters in distress.

This cultural attitude principally determines women's status in society, leaving them completely dependent on the whims of men, be it father, brother, husband or son. Despite the tradition of bride price, the money is controlled by men. If there is any conflict in the family, the woman has few options to moving out as she owns neither land nor house. She must either find another man to give her shelter or go back to her parents' or brother's house. This is in a context where domestic violence, alcoholism, bigamy and desertion are common among men.

Moreover, the traditional division of labour between men and women binds her more intimately to the natural resource base. It is she who pays the price for environmental degradation by having to walk longer distances in search of firewood, fodder and water. Thus, while having an intrinsic interest in maintaining a healthy natural environment, she finds herself devoid of control over decision-making related to management of a basic resource like land because its ownership vests with men.

Evolution of Women's Involvement in PEDO's Environmental Rehabilitation Programme

PEDO's Women's Programme and its Environmental Programme have evolved over a number of years through a fairly rich and flexible process entailing a lot of learning and questioning. Neither followed any predefined targets or project structure and women's involvement in the environmental programme got defined through problems experienced in working only with women. Thus, over time, both programmes got considerably modified, one influencing the other. Here, it is useful to briefly trace the evolution of both programmes through different phases.

The Origins of PEDO's Work with Local Women

In 1980, one of the first projects taken up by PEDO (then the SWRC) was to train selected groups of women in *durree*-making and handloom weaving.

When the improved *chulha* programme, based on training selected village women, was started in 1983, some of the women who had already been associated with handloom training came forward to be trained as *chulha mistris*. By 1986, the number of trained *chulha mistris* and supervisors had increased to thirty. All of them attended monthly meetings at Mada to report on progress and problems. As they were responsible for providing *chulha* users regular follow-up support, each one of them had established contact with 102 local women in their respective areas. It was through this extended network of contact with local women that the organisation started gaining a better understanding of their problems. Many of the women *chulha mistris* and supervisors were themselves affected by some of the typical problems of women in the area. Inevitably, their personal crises affected their work and the programme repeatedly lost trained women.

Simultaneously, with the commencement of three years of drought (1985 to 1987), the near total collapse of the natural resource based subsistence economy started becoming evident. Its impact on all local people was terrible but once again, women appeared to be the worst victims.

A major casualty was health. Three years of crop failures reduced food availability and nutritional diversity drastically with a consequent increase in infections and disease. While women were equally, if not more, affected by this, their burden of caring for the sick members of the family also increased.

Parallel to this, anxiety and physical workloads increased. With migration of men to other areas in search of wages, women were left behind to shoulder men's responsibilities. Then, there was the added fear that the husband might pick up another woman while away and the possibility of physical abuse by other men due to the husband's absence.

The situation led to a serious review and reflection within PEDO which laid the foundations of its present women's and environmental programmes. Women's problems—made evident by the *chulha* programme—led PEDO to decide to formulate a broader women's programme to address them effectively.

It was clear that none of these problems could be dealt with at the individual level. Organised and collective action was a prerequisite. This meant seeking women's reflection and involvement in articulating dominant issues confronting them and defining appropriate strategies of action. For this, the *chulha mistris'* wide network of contacts with local women could help reach out to a still larger number of women. With this objective in mind, a large *mela* of women was organised in Mada in April 1987.

The response to the *mela*, despite several problems, was overwhelming. During the three days (13–15 April 1987), 3000 women, mostly rural, instead of the planned 1500, participated! Although the majority of them were local, women from 41 groups and organisations from different parts of the country also came. Besides endless, uninhibited singing, dancing and playing games together, the women shared their experiences, joys and sorrows. For most of them, this was the first time in their lives that they had been away from their domestic chores or travelled so far away from home.

The major problems articulated by the women during the *mela* fell into two main groups. The first essentially comprised economic problems. These could be linked to the virtual collapse of the agrarian economy during the prevailing drought. Topmost among the problems affecting women more acutely were the scarcity of fuel, fodder and water, forcing them to walk longer and longer distances in search of them. Due to inadequate production in the subsistence sector, there were the related problems of unemployment, increased dependence on wages from drought relief works, corruption in the execution of these works and in the management of government-run ration shops, and shortage of cash for emergencies like illness, leading to increased dependence on exploitative money-lenders.

The second group of problems were social in nature: alcoholism among men, domestic violence, desertion, bigamy and the general insecurity in women's lives. The women also complained about the poor functioning of the government health and educational services.

The *mela* generated tremendous enthusiasm and represented a watershed in the organisation's rapport with women of the block. Several misconceptions about PEDO were removed and, for the first time, groups of village women started approaching PEDO to assist them in solving some of their problems.

Due to considerable discussion on the causes and impact of the prevailing drought, women of several villages who had participated in the *mela*, asked for fuel, fodder and fruit plants for planting on their private lands. Women of three villages asked for assistance in rehabilitating their totally degraded village common lands. Follow-up meetings after the *mela* were held with hundreds of women who attended it. Through these, PEDO took its first step in involving women in developing both private and common wastelands.

The Initial Experience of Women's Involvement in Developing Private and Common Wastelands

After the *mela,* the women staff were asked to prepare lists of demands for plants from their respective field areas. It was decided that PEDO would supply free saplings of the species of the women's choice on the condition that pits for the plants were dug in advance and on their giving an undertaking to protect the plants after planting. A demand for 30,000 saplings was received from women of about 60 villages.

However, after the first monsoon rain of 1987 when the distribution of saplings was started, the demand for plants from women shot up phenomenally. Ultimately, a total of almost 164,000 saplings were distributed to 602 women spread over 68 villages! In addition, 6,000 grafted plants of mango and lemon were distributed. PEDO's staff had to overstretch themselves to meet this demand. All this resulted in inevitable confusion with plants reaching many women too late, inadequate supervision of planting, damage to plants while transporting them to far-flung villages and many women not getting the species of their choice.

Inevitably, mortality among these plants was high. The almost complete failure of the monsoons in 1987 did not help matters. Many women expressed their inability to protect the plants due to having to leave home daily in search of wage work to survive. Hungry cattle roaming in search of anything green made protection even more difficult than usual.

However, the experience made one thing clear—that the drought had made women realise the importance of re-establishing vegetative cover on their denuded lands and, if provided with some guidance and support, they would come forward in large numbers to plant trees to reduce their problems of fuel and fodder scarcity.

The three months between the women's *mela* and the planting season was too short a time to plan PEDO's intervention systematically. For the organisation's women staff, this work was new and there had been no time to provide them even preliminary training in plant handling and planting techniques. Their responsibilities for supervision after planting had not been defined clearly. In any case, due to the unexpected upsurge in demand at the last minute, they lost track of where many of the plants had been planted. The dispersed settlement pattern of the local *adivasis* made it extremely difficult to check on plants planted all over the area.

Although PEDO had started planting on the fairly abundantly available degraded common lands in 1985, prior to 1987 its main preoccupation was to make green again the barren lands. It took the responsibility for protection itself and excepting for some dialogue with a few leaders, there was limited community involvement.

In 1987, discussions on evironmental issues during the *mela* prompted women of Ditkon Ka Vela, Bortalav and Baletighati villages to request PEDO's help in rehabilitating their degraded common lands. PEDO agreed to support the efforts of the women's groups of the first two villages that year. By this time, the near impossibility of effective protection of plantations without the community accepting responsibility for it had made PEDO realise its importance at all stages of work.

It was in Ditkon Ka Vela and Bortalav that, for the first time, the condition of right-holding families accepting responsibility for protection was made a precondition for the organisation providing assistance for developing common lands. Although initially, people of both the villages resisted this concept, the women's group of Ditkon Ka Vela came forward and agreed to contribute a fixed sum from their wages towards building up a village common fund from which payments for protection would be made. The men agreed later.

Learning from this experience on private and common lands, planning for the 1988 monsoon was started early. To begin with, a participatory evaluation of work done during 1987 was conducted by PEDO's own staff. In November 1987, an eight-day training programme was conducted for PEDO's women staff. An important component of this was to get the women to reflect upon and analyse the environmental changes the area had witnessed in recent years and their causes and impact on their day-to-day lives. This proved to be an eye-opening exercise for the women. They had never looked at the totality of the problem in this manner before. For the first time, they consciously understood the significance of trees and forests for their survival. Whether it was a decrease in soil productivity, reduced and uncertain rainfall, declining water-table, drying up of perennial water sources, scarcity of fuel, fodder and nutritious foods or increased health problems, they could trace the cause to the destruction of forests.

This was followed by getting four groups of women to do a planning exercise on the number of trees and shrubs a particular member of the group required for meeting her own firewood, fodder, construction material and other subsistence needs. This generated tremendous interest and excitement among the women and provided them a concrete basis for discussing action strategies with other women. For example, the women calculated that for supporting a cow or a buffalo entirely on tree leaf fodder for the whole year, 90 trees would be required for just one animal. For a goat, 50 trees would be needed. A list of 21 local species suitable for fodder was also generated. Similar calculations were made for the timber required for building a two-roomed house. The women were quite amazed at the number of trees needed for meeting their basic needs.

Similarly, on the basis of the previous year's experience, basic principles for working with groups on common lands were evolved.

Tensions Generated by the Focus on Women

However, at this stage it started becoming evident that the focus on working with women was beginning to generate tension and hostility not only among male villagers but also among PEDO's male staff. The male staff started questioning the objectives of the women's programme. They could see few tangible outputs from it unlike the handpump installation, well-deepening or contour-terracing programmes with which they themselves were involved. The women's programme, with its emphasis on group building and eliciting women's total participation, seemed to them to consist of endless meetings with no tangible results. They complained that when they were asked to intimate village women about a meeting, the village men wanted to know why they were not being invited. Reports also started coming in of some of the village women being beaten by their husbands when they returned empty-handed from a women's meeting. Due to the prolonged drought, everyone was desperate for getting even daily wage work. Some men had been permitting their wives to attend the women's meetings in the expectation that the organisation would provide some employment to them. When they returned empty-handed, the men's frustration found release in wife-beating.

We also started realising that although women were the greatest sufferers of environmental degradation and could take the lead in remedial action, the socio-cultural context made them dependent on men's support for taking any initiative. Almost all the private land is owned by men. Very few women are strong enough to assert themselves in deciding how this land should be used. Even some of the women staff, despite their conviction and excitement about the need for better management of their lands, felt unable to see their ideas through unless their husbands agreed.

Similarly, the women staff needed the male staff's cooperation and support while working in villages. If the male staff were not convinced about the validity of what the women staff were trying to do, the women staff had to face yet another obstacle in their already difficult task.

All this led to the decision to involve men in the process of change. The PEDO's objective was not to create yet another divide between men and women. Instead, the aim was to strengthen women's capabilities to deal with their problems and fight their oppression both individually and collectively. If men were made participants in the process of reflecting upon how women's specific burdens had increased, they might cooperate in finding ways of reducing them. But if, instead, they felt defensive and threatened, it was likely to result in increased oppression of women in a situation where women were ill-equipped to deal with it. Unless accompanied by a powerful social movement, individual women cannot suddenly challenge men's traditional supremacy in decision-making related to land.

Working with Joint Groups of Men and Women on Common Lands

Taking these developments into account, it was decided to modify the approach and involve men in the programme. To begin with, during a 10-day training-cum-evaluation with PEDO's entire programme staff in January 1988, the impact of environmental degradation on women and the role that women could play in an environmental reconstruction programme was made an important aspect of discussion and reflection. The male staff's response to this was heartening. They themselves suggested that in all future meetings with villagers, at least one woman staff member should be present to facilitate women's participation. The male staff also highlighted the problems of mobility faced by the women staff and supported the latter's demand for being provided bicycles and/or mopeds. To reduce village men's suspicions and hostility to separate meetings with women, it was decided that, to begin with, joint meetings of men and women would be called in every village. As women's participation in such joint meetings is almost always limited, the need for having separate meetings with women would be explained.

From early 1988 onwards, the broad objectives of PEDO's environmental programme became more clearly defined as follows:

1. To reduce the people's almost complete dependence on unreliable, rainfed agriculture by increasing the diversity of crops to include economically useful trees and shrubs;

2. to find practical alternatives to uncontrolled grazing;
3. to find practical methods for harvesting every drop of rain water;
4. to divide workable community managed systems of protecting and rehabilitating degraded common lands to increase their productivity and to work towards women's equal participation in these.

To begin with, equal emphasis was given to promoting the development of both private and common wastelands. With experience, however, priority started being given to common lands. Initially, the staff felt that it would be easier to motivate people to work on their private lands compared to developing a common resource. Within a year of trying to work with hundreds of individuals on their landholdings scattered over a vast area and dealing with their suspicions and reluctance to replant on land they had cleared with much effort, the staff started feeling that it was both easier and more desirable to work on common lands. Questions of equity were also involved. While private wasteland development would only benefit individual owners, development of common lands would assure benefits to all members of the community. This necessitated developing the staff's skills to work with groups and resolving conflicts within groups. Periodic follow-up and review workshops with the staff have been held for this purpose.

There are large chunks of degraded common lands in the block. These are either *chernot* lands or revenue wastelands. Although management of the *chernot* lands has been vested in the *gram panchayats,* residents of particular villages or hamlets have clear, *de facto* grazing rights in specific pieces of the lands. Stone protection walls have been built around several *chernot* lands by *gram panchayats* with drought relief funds.

While defining clear principles for PEDO to work on common lands in early 1988, it was decided to build upon the existing group identity of sharing traditional rights in them. Thus, *all* right-holding families were to be involved in the group. To make every family feel equally responsible for protection, each family was to be assured a right to an equal share of the produce. Each group had to make a prior commitment to protect the plantation before PEDO was to get involved. Day-to-day management decisions had to be taken by a Managing Committee (MC) elected by the whole group. There had to be at least an equal representation of women on the MC which was to meet regularly every month. Regular meetings of the entire group were also to be held and every effort made to encourage women's participation.

Training Women for Seed Collection

To ensure availability of good quality seeds for its growing planting programme, especially of promising local species PEDO organised a training programme in seed collection and storage in May 1988. Thirty women from 13 villages were taken to visit the Seed Bank of the Jawaja Project in Beawar. It was felt that imparting technical knowhow about seed collection to local women would not only enhance their status within the community, but would also enable them to combine their fuel-gathering and cattle-grazing activities with earning some income. More importantly, it could stimulate women's participation in monthly meetings at the seed collection centres and lay the foundation for developing strong women's groups.

Village meetings held after the seed collection training clearly indicated that as a sub-activity of the environmental programme, seed collection by women held tremendous potential. It could provide a basis for the formation of women's groups which could hold their own even in mixed meetings related to the plantations. Unfortunately, however, due to inadequate women staff and lack of sustained follow-up of the women's seed collection groups, these have not been fully integrated into the women's programme. Women engaged in seed collection do attend plantation meetings more regularly and enjoy a better rapport with the organisation. In some villages, they have also joined the women's saving groups. However, the organisation needs to provide them more sustained and systematic support to build up their full potential.

Review of Women's Participation in Joint Groups and Charting a New Course of Action

As already mentioned, due to the evident hostility among men to PEDO working separately with women, it was decided to work with joint groups of men and women on common lands. In any case, as traditional use rights to common lands are family-based, it was felt that working on them only with women would generate avoidable discord. At the same time, due to women's secondary status even in *adivasi* society, there was considerable concern about whether women would have any effective participation in groups in which men were also members. All that the organisation could do to facilitate women's participation was to make it a precondition for working with any community and ensure the availability of women staff during village meetings to encourage women to come out of their shells. But even this cannot be always ensured either due to inadequacy of women staff or some of them being unable to provide the continuous and regular support required due to instability and personal problems in their own lives.

Further, equal participation by women was being sought in a cultural context where there is a strong surviving tradition of men getting together twice every month on the eleventh day of the lunar cycle to discuss village affairs. Women are completely excluded from this traditional village forum. As a consequence, the men are always better informed and used to taking decisions related to village affairs. Also, despite winning over the male staff to the cause of facilitating women's involvement, and providing them periodic training inputs for developing the skills for it, it is too easy for the male staff to revert to their traditional dominating or patronising role towards women. After all, they come from the same cultural background and age-old traditions and attitudes cannot be changed overnight.

As members of the resource team assisting PEDO to evolve its programmes, we were unsure whether the decision to work with combined groups of men and women was really a positive step or whether it was effectively going to diminish even the limited women's involvement which had become evident while working exclusively with women.

The performance of this approach was reviewed during a training programme for women staff in September 1988 while discussing other issues of concern to women. The consensus that emerged was that, although women were the worst sufferers of environmental degradation, their lack of land ownership and traditional male domination effectively inhibited them from playing a leading role in plantation-related decision-making, except in a few villages. Joint meetings of men and women, while successful in reducing male hostility and securing their cooperation, tended to diminish free expression and articulation by women. The need to create a separate forum for women in which they could express their views and concerns, uninhibited by the presence of men, was strongly felt.

At the same time, the 1988 monsoon was an exceptionally good one after four to five years of poor rains. This, in itself, had created a new situation. Large numbers of cattle had died during the drought years. However, due to the good monsoon of 1988, there was abundant grass. Within a span of a few months, the situation of fodder scarcity had changed to one of fodder surplus. Instead of problems of fodder and water, the women were more preoccupied with getting credit for buying grain seeds for sowing. All the stored grain had been eaten up during the years of scarcity. Also, due to a dramatic increase in illnesses during the drought years, there had been an increased need for emergency loans. The only accessible credit source was the local money-lender. During this training programme, women talked about the extreme exploitation by local money-lenders. They responded enthusiastically to the idea of starting women's savings groups as a long-term measure for getting out of the clutches of money-lenders.

The growth and development of about 30 women's savings groups since then has been a remarkable phenomenon and demands for assistance in forming new groups are continuing to come in. Surprisingly, there has been next to no male hostility to this activity being undertaken exclusively by women. The monthly meetings of these groups have also provided the much needed forum for women to get together regularly for a non-threatening, non-controversial purpose. While some of the groups have kept themselves confined to the savings activity, many have taken up several other issues of concern to women. The existence of active women's savings groups in villages, where plantations on common lands have been taken up, has generally facilitated more active participation by local women in their management. Three-day awareness generation 'camps' for women have been held in some of the villages having savings groups and inter-group visits and discussions have been organised to facilitate learning from each other's experiences. In the process, group members have gained confidence in articulating their problems, acting on them and being able to talk even in gatherings where men are present.

A three-day get-together of more than 400 women belonging to various savings and plantation groups was held in May 1990. During the meeting, it became evident that local women's groups have become dynamic entities and a force to reckon with. On several occasions, women group members have demonstrated greater unity and courage than the men. In many villages, women's groups now command a grudging respect from other villagers.

Case Studies of Women's Participation in Development of Common Lands in Three Villages

Till 1989, PEDO had assisted 29 village groups to rehabilitate a total of 1,304 hectares of their common lands. The organisation has stuck to its precondition of women's participation in all the groups formed since 1988. Women's actual and active participation, however, has been highly variable. It is difficult to isolate all the factors which have influenced this. In this section, we look at the performance of three groups in greater detail in terms of women's role and participation at different stages. Two of them have had, and continue to have, strong participation by women while in the third one, it has been limited. All the three plantations are at least two years old.

Ditkon Ka Vela

The common land plantation in this village was initiated in 1987 and was the first to be taken up by PEDO at the request of village women. Six women from the village had participated in the women's *mela* in April 1987. Two of them, Galali and Ratni, took the initiative and have continued to play leadership roles. At that time, the goals of PEDO's environmental and women's programmes were being redefined and both the staff and the village women were full of enthusiasm generated by the *mela*. PEDO's principles for working on common lands had not yet been worked out clearly. Ditkon Ka Vela and the adjoining village of Bortalav served as experimental cases for trying out the principles: villagers accepting collective responsibility for protection and open decision-making in village meetings. The response of the village women here served as an inspiration for other villagers to come forward.

Socio-economic structure of the village

Ditkon Ka Vela has about 60 families, all of whom are *adivasis*. All families are dependent on rainfed agriculture on small land-holdings. There are only seven wells which are used for irrigation. Because of this, most families can raise only one crop in a year and even that fails if rainfall is either inadequate or untimely. About 40 per cent of the families supplement their income by working in nearby stone quarries and members of about 30 per cent of the families seasonally migrate to Gujarat for wage work. The nearest road is seven kilometres away and virtually no government services reach the village. Due to extensive deforestation all around and the consequent fodder shortage, there are only 28 buffaloes in the village although all families own some goats or sheep. Women have to walk at least two to three kilometres to fetch firewood from the nearest forest although most families have small protected forest patches on their private land to meet some of their fuel and fodder requirements.

The common land plantation

Plantation has been done on 14 acres of *chernot* land on which all villagers have traditional grazing rights. Due to totally unregulated exploitation, nothing except a few stunted trees were left on the land. A family from the adjoining village had started encroaching on the land which seems to have first motivated the villagers to do something to save their common resource. In 1986, the *gram panchayat* built a stone boundary wall around the land and got planting done with government drought relief funds. None of the plants survived, the boundary wall collapsed at several points and the villagers did not receive full payment for the work they had done on the land.

It was to replant on this land that the women approached PEDO in April 1987. It should be recalled that 1987 was one of the worst drought years of this century. Poor monsoons during the previous three years had further impoverished the people and everyone was desperately looking for daily wage work on government drought relief programmes. These were riddled with corruption and exploitation. Village leaders were generally giving disproportionate employment to their own family members on such works. Some of them did no work at all or lorded over the others by becoming 'mates' of labour gangs. There was little accountability for the actual work done and no discussion about the future management of afforestation projects.

Women's role in the process

It is in the above context that women's role in the common land plantation of this village needs to be traced. Some of the landmarks in the process have been:

—In the initial meetings with PEDO's staff, more women than men took part. Although everyone had immediate interest in getting daily wage employment through the project, women were more willing to discuss long-term protection and management questions following discussions during the *mela*.

—When PEDO insisted on the condition that all the families give an advance undertaking to protect the plantation after the planting was over, the men, led by the *gameti,* were unwilling to accept this. It is the women who agreed and gave an undertaking to contribute a small amount every month to pay for a watchperson's salary for protection.

—In the initial Managing Committee (MC) elected by the villagers in July 1987, out of 17 members as many as ten were women. However, till the following year, the *gameti* kept objecting to women being made members of the MC. During a meeting in June 1991, the number of members was reduced to nine out of which six are women and three are men.

—There was always a much higher percentage of women than men doing the earthwork and planting. Women's productivity was always higher than that of men. When it was found that the cost of the physical work was too high due to payment being made on the basis of daily wages instead of output, women willingly accepted self-monitoring of output to ensure the minimum productivity required.

—It was the women who took a stand against the family members of local leaders receiving disproportionate employment on the plantation. As a result, the principle of employing only one person from each family was adopted. This has been used in all subsequent plantations.

—In February 1988, seven goats belonging to three villagers grazed on many of the plants. On the insistence of the women, all three were made to pay a fine of a total of Rs. 130 as per the rules formulated by the MC. This sent the signal that such violations would not be tolerated in the future. Because of this, protection has been excellent since then with a plant survival of 90 to 95 per cent. Today, the originally barren land looks like a healthy young forest with trees of some species 10 to 15 feet tall and a good grass cover on the ground.

—In 1987, the first crop of naturally regenerated grass was harvested from the plantation by voluntary labour. One person from each family came to cut it on a day fixed by the group's MC. Most of those who came to harvest the grass were women. During 1988, the increased grass yield was harvested with voluntary labour and sold for Rs. 800 which was deposited in the common fund. There was no demand for grass in the village due to heavy cattle mortality during the drought years. The 1989 grass yield was sold for Rs. 500. Although the grass output increased between 1988 and 1989, the price fell due to surplus availability of grass.

—In November 1988, a women's savings group was started in the village. Sixteen women became members and saved a total of Rs. 1299 within one year. Regular monthly meetings of this group have further strengthened women's participation in the management of the common land plantation. The women's group has also initiated action against the malpractices of LAMPS (Large Area Multipurpose Societies set up by the government for direct marketing of minor forest produce in the tribal areas) and local money-lenders.

—The women have taken a clear decision that when fuelwood becomes available from the plantation, it will *not* be sold to outsiders as women have to labour hard to collect it from distant forests.

Initially, whenever an important decision had to be taken during a village meeting, the men would move aside to discuss the matter among themselves and return to the gathering to announce their decision. During our field visit for this case study, in a similar situation, the women also went aside for a similar discussion! In any case, 21 women attended the meeting compared to only nine men. This is akin to the beginning of a new cultural tradition. When asked, the women said that till the women's *mela* in April 1987, women had never participated in village decisions. Only the men took these decisions during their *agyaras* meetings. They are happy to be better informed and consider their involvement in village affairs to be a positive development.

Several women of Ditkon Ka Vela have been exposed to the activities of other women's groups through study tours, inter-village visits, *melas* and training programmes. The benefit they have derived from this is reflected in their increased level of confidence and articulation and in their growing group identity.

Unresolved problems

The village group suffers from the following, yet unresolved, problems:

—A sub-group of about 15 families has not been participating much in the village group;

—the common fund of the group has no funds left for paying a *chowkidar* and little monetary return can be expected from the plantation in the near future. However, during our recent meeting with the group, the women once again made a commitment to raise the money through monthly contributions;

—PEDO has inadvertently assisted a couple to build up a power base for themselves. They make it appear as if they are responsible for getting benefits for the village from the organisation. The villagers have not yet gathered the courage to challenge them openly to check the unfair advantages they take due to this. The fact that the villagers consider the woman to be a *dakin* wielding evil powers makes them reluctant to confront her. The couple's behaviour is becoming a threat to the unity and dynamism of the group.

Bhimsor

This village has 67 families, all of whom are *adivasis*. Compared to other villages, people here have smaller landholdings of only two to three *bighas*. The land is not very fertile either. Most families keep only goats as the surrounding hills are badly denuded with nothing other than thorny bushes left on them. As a consequence, there is an acute shortage of both fuel and fodder. Due to the limited income from both agriculture and livestock, several men of the village have either migrated to other places for work or work in nearby stone quarries. About eight to 10 men have salaried jobs working as drivers or conductors.

The village is divided into two opposing factions over whether the embankment of a village pond should be raised or not. If the embankment is raised, fertile agricultural land of more than half the families of the village will get submerged. These families have been opposing the project for more than 10 years. Apparently, the women of this faction have been in the forefront of the struggle all this time. They have instead been suggesting that the embankment of the other village pond be raised. They claim that this will help irrigate the lands of a much larger number of families of the village. The *panchayat* member from Bhimsor belongs to the faction supporting the disputed project. His primary motive seems to be to denigrate the dynamic women's group opposing the project. Apparently, on several occasions, he has commented that the women in the group are getting too big for their boots and unless put in their place, men will soon have to start wearing bangles! The most recent confrontation over the pond took place in June 1991. The women's group, supported by their men and children, camped near the pond for ten days and nights, refusing to be intimidated even by threats from policemen. They successfully checked another attempt by the *panchayat* to forcibly raise the embankment and forced the *sarpanch* to give in writing that the work will be suspended till both factions reach a settlement.

With many men living and/or working outside the village, women have to take care of the subsistence sector of the village economy. Because of this, they tend to be more articulate and outgoing. Due to the extreme shortage of fuel and fodder, they have a strong interest in rehabilitating degraded lands. And possibly due to the women of the majority faction having been involved in the dispute over the pond for 10 years, they have developed a strong group identity. Today, the Bhimsor group is among the strongest women's groups PEDO is working with, not only in relation to wasteland development but several other activities as well.

PEDO's contact with Bhimsor was established in 1984 when the villagers requested its assistance in starting a primary school. From 1984 to 1987, the organisation implemented well-deepening, soil conservation and hand pump installation programmes in the village which strengthened its contact with the villagers.

Women's participation in wasteland development

About 10 to 12 women from Bhimsor attended the women's *mela* at Mada in April 1987. Following this, they expressed interest in planting trees on their private lands and received free saplings the same year.

In 1988, plantation on private wastelands in the village was done more systematically and women played a leading role. Their major constraint was their small wasteland holdings which they also use for raising grass during the monsoons. After the rains, the women were reluctant to do hoeing/weeding of the plants as they wanted the grass as well. As a consequence, despite good protection, plant survival was only about 40 per cent.

In November 1988, a women's savings group was formed in the village. By October 1989, the group had saved almost Rs. 2,000 in their collective account.

In February 1989, a three-day village camp was organised in Bhimsor. This helped in building up an extremely cohesive and dynamic women's group. The group has since been in the forefront of the struggle against the exploitative practices of LAMPS officials, and has received support from the village men. The group also wants to experiment with starting a village grain shop to eliminate exploitation by middlemen.

Following the discussion on the state of the environment during the village camp the women asked PEDO for assistance in rehabilitating 60 *bighas* of their degraded common grazing land. They themselves obtained a map of the land from the *patwari*. They also got a resolution signed by six members of the *gram panchayat* that only Bhimsor residents have rights in the land and, therefore, to the produce from the plantation. About 14 to 15 village meetings were held prior to starting earthwork on the land. The villagers accepted complete responsibility for protection and agreed to deposit one out of every 15 days' wages into a common fund to pay for a watchperson. A Managing Committee of six members was elected unanimously. Out of these four are women and two men. Two women have been authorised to operate the group's account. The common fund has Rs. 2600 in it at present. If this money runs out, the women say they will protect their plantation with voluntary labour. A sum of Rs. 80 has been recovered as fine from persons whose goats had entered the plantation. The group has made a rule that anyone's cattle caught inside the plantation will be tied up by the *chowkidar*. If the owner does not take it away within twenty-four hours by paying the fine, it will be auctioned and the money deposited in the common fund.

The women have also repaired damaged parts of the stone boundary wall and placed thorny bushes on top of it through voluntary labour. They also dibbled mango and *mahua* seeds in 1989 and in June 1990, planted branch cuttings of two to three types of fencing shrubs along the boundary wall to strengthen protection.

Over 95 per cent of the earthwork on the land was done by women. In fact, only two men worked on the plantation. All the rest were women.

The planting itself was done in 1989. After just a few months of protection, some grass had come up through natural regeneration. The women harvested the grass through voluntary labour and two women from the village bought it for Rs. 50. This year, when there is likely to be more grass, the women are planning to charge Rs. 3 to Rs. 5 per day person from members of the group to harvest it. This is one of the few villages where, despite the present general grass surplus, some women have had to buy grass and even the limited grass available from the plantation has been bought within the village.

The other heartening aspect of Bhimsor's common land is that within barely one year of protection, a wide variety of local species have started regenerating naturally. A lot of rootstock was surviving under the barren soil.

This is an exceptional group which, effectively, is a women's plantation group despite being called a mixed group. During our meeting with them, there were seventeen women and only two men present. The men do not seem to object to the women taking part in project activities, visiting government offices or taking initiative on their own. Only two or three men accompany the women during such ventures. Some of the women have developed strong leadership qualities which, they feel, have improved since the camp held in the village in February 1989. The commitment and involvement of both PEDO's staff members (a man and a woman team) working in this field centre have also contributed to the development of this group.

Paderi

The struggles and initiatives of the Bhimsor group have inspired the emergence of a new, equally dynamic group this year in the adjoining village of Paderi. Two women of Paderi have their parental homes in Bhimsor. While visiting their relatives, they got acquainted with the activities of the Bhimsor group and asked PEDO for assistance for a similar plantation in Paderi. Women there also face acute scarcity of fodder and firewood as forests have been completely decimated all around the village. Many families now use dung as cooking fuel or have to buy a bundle of firewood for Rs. 10 to Rs. 12. Shortage of firewood is so acute in this area that many families are now forced to bury their dead instead of cremating them.

These two or three women have identified 352 *bighas* of revenue wasteland near the village on which the villagers claim they have traditional grazing rights. The *chernot* land of the village could not be taken up for plantation as several families have encroached on it. A few families were doing illegal stone quarrying even on the revenue land on which work has now been started.

The women who took the leadership in getting this plantation going went to the *tehsil* office four times on their own to complain against the illegal stone mining. With great difficulty, they managed to extract the revenue map of the land from the *patwari* who was in league with those engaged in quarrying. They have also obtained a resolution from the *panchayat* signed by six members that Paderi residents will have sole rights to the produce from the plantation. The women received little support from the men and had to face the hostility of the families forced to stop illegal stone quarrying. The commitment of one of the women leaders is so high that she says she is prepared to die fighting against those opposing the project. Despite such inspired struggle, only 175 *bighas* have become available for the plantation. The rest of the land had to be left out due to pressure of vested interests.

However, with the completion of the stone boundary around the land and the commencement of earthwork, support for the leading women has grown and the staff member in charge of the area has won over the displaced quarrying families by involving them in raising nurseries and providing them wage work on the land.

Paderi is a larger village with a total of about 132 families out of which about a hundred are *adivasis* and the rest are Patels. Its problems are by no means over as the underlying conflicts within the village are bound to surface again. Consolidating the process of group formation is likely to take time. But the women who took the initiative are likely to get support from other village women as the plantation will reduce their work of collecting fuel and fodder.

Wagbole

Although a mixed plantation group has existed in Wagbole since 1988, and PEDO's condition of women's participation was equally insisted upon, women's active involvement in this group has been marginal. The process of group formation, the socio-economic structure of the village and its internal dynamics are examined here to understand why this has been so.

The village and its economy

Wagbole, like the other villages, is spread over small hillocks but here, deforestation has not been so acute. Nearby forests are in relatively better condition and all the 42 *adivasi* families of the village have small protected forest patches even on their private lands. The average landholding is a bit bigger than in other villages (about 10 *bighas.*). Due to comparatively less soil erosion, both the agricultural and common lands have better soils and higher productivity.

The main occupation of the households is rainfed agriculture from which they can get only one crop in a year. Only five or six families have wells for irrigation. Even so, agricultural production in the village is relatively better. Several village men go outside the village for daily wage work and some men have migrated to Gujarat.

The nearest road from the village is five kilometres away because of which the villagers remain relatively cut off from the outside world. Hardly any government services have reached the village and the sick have to be carried long distances for treatment.

Due to the healthier subsistence economy of the village, family labour, particularly women's labour, is required to undertake cultivation work and livestock rearing. As the village is fairly well-endowed with its own fuelwood and fodder resources, the women do not have to walk long distances in search of them. Possibly because of this. the village women have remained almost unexposed to the outside world and men are resistant to letting them go out. The village men are relatively more conservative and fall easy prey to rumours that outsiders want to destroy their religion by converting them to Christianity.

PEDO first established contact with the village in 1984 when a primary school was started in an adjoining village. During 1986–87, well-deepening and contour-bunding were done in the village.

The common land plantation process

In this case, it was the men from Wagbole who approached PEDO to assist them in planting on 60 *bighas* of their common grazing land. The *panchayat* had already built a stone boundary on it and planted there once. The wall had already collapsed in places and none of the plants survived. The villagers had asked the Forest Department to replant on the land but the Department had turned down their request saying that the area was too small.

Then, impressed with the progress of the plantation in the neighbouring village of Ditkon Ka Vela, the men approached PEDO for similar assistance. PEDO clarified the conditions on which it works on such lands. After considerable initial resistance, the men agreed to accept responsibility for protection and to put aside one day's wages out of every 15 days' work done on the plantation into a common protection fund. Planting was done in 1988. The primary motivation of the men at that time was to generate daily wage employment for the villagers.

Although women were called to the initial meetings, it always proved difficult to involve them in the discussions. One male village leader tends to dominate all the time and even other men are unable to say much in his presence. The women get little chance to open their mouths in such mixed meetings. A few separate meetings with the women were held initially to build up their confidence. However, these could not be sustained for long due to inadequate women staff. There has also been a high turnover of women staff at this field centre and with most of them being new, they have been unable to deal with the situation.

The men accepted having three women on the six-member Managing Committee appointed for the plantation. However, it is unclear whether they were selected by the women themselves or their names were simply put on the committee to satisfy the organisation's condition.

During the execution of physical works on the land, more women than men worked there on daily wages. At that time, it used to be relatively easy to hold impromptu meetings with the women on the paltnation site itself.

Since completion of the work, attendance at monthly meetings of the group has generally been poor. Only the male committee members tend to attend these and only two or three women participate.

Facilitation of women's participation

In terms of management of Wagbole's plantation, there is little one can complain about. The villagers have protected it extremely well. Besides good plant survival, there has been excellent natural regeneration of several local species. Grass production has also increased dramatically. Prior to the planting, the land was used for open grazing and there was little vegetation on it.

The watchman is changed by the group every month to distribute the benefit, of a salary from the common fund, among all village families. In 1988, every family was invited to harvest grass on the payment of Rs. 5 per head. In 1989, due to abundance of grass, there was less demand for it. But rats proliferating in the grass were damaging the plants and there was danger of the dry grass catching fire if not harvested. Every family was asked to harvest it on the reduced payment of Rs. 2.50 per head. There was also about 40 kilograms output of castor seeds from the plantation. Some of these were kept as seed for the following year and the rest sold and the money deposited in the common fund.

However, as noted earlier, women's participation has been limited. The organisation attempted several methods for increasing it. About four to five women from the village had attended the women's *mela* of April 1987. They were certainly very enthused by it and at least one or two of them have continued to be relatively more active. Then, an attempt was made to hold separate meetings with women while work on the land was going on. These also made the women more aware about how the plantation was to be managed by villagers themselves and the importance of their taking an active part in it. This was followed by training two village women in seed collection for PEDO's Seed Bank. Regular contact was established with them and they could earn some income from the activity.

Despite all this, the women have not felt empowered enough to start asserting their views more meaningfully. Perhaps because of the privately owned good forest paches, the women of Wagbole have never felt the negative impact of environmental degradation in the way women of other villages have and are therefore not adequately motivated to participate in plantation-related meetings. In fact, in late 1989, PEDO had organised an environment camp in the village. During the camp, it was planned to discuss future management of the plantation and systems for harvesting, distributing and using produce like leaf fodder, firewood, and poles from it in the future. Surprisingly, no one from the village came for the camp. After a lot of persuasion, a few men came but it appeared that the women were either not permitted or were unwilling to come at all. Later, it was found that a rumour had spread that the camp would be used to forcibly convert the villagers to Christianity! Similarly, in the gathering of members of women's groups in May 1990, only one woman from Wagbole participated. Apparently, rumours about women being taken away by strange men were still doing the rounds and the village men were too suspicious to let their women participate in the event. The absence of an experienced woman staff member working at this Field Centre (FC) on a sustained basis is probably also responsible for the women's continuing lack of faith in the organisation.

Only recently, in June 1990, seven women of the village have decided to start their own savings group. If the membership of this group increases and it gets consolidated, possibly its regular monthly meetings will become an effective forum for reducing the women's isolation. That might even lead to more active participation by the women in the management of the village common land.

Wagbole's experience makes it clear that women's participation cannot be ensured simply by making rules insisting that they do so! This is particularly so in a social setting where women's traditional roles are staunchly guarded by men.

THE SITUATION TODAY AND DIRECTIONS FOR THE FUTURE: WOMEN'S PARTICIPATION IN OTHER GROUPS

There is considerable variation in women's participation in other common land plantation groups with which PEDO is working. First, it is useful to list the commonalities in all the groups:

—At least 75 per cent (in some cases 90 to 100 per cent) of the labour for earthwork and planting has been women's labour;

—where grass has already been harvested, whether through voluntary labour or on the basis of payment, most of the grass-cutting has been done by women;

—in the future, when firewood starts becoming available for harvesting, even this is likely to be cut primarily by women;

—where the *initiative* for undertaking the plantation was taken by women (e.g., Ditkon Ka Vela, Bortalav, Baletighati, and Bhimsor), they have continued to play a leadership role in subsequent management- and protection-related decisions. They have often made commitments (such as making contributions to the common fund for protection), which men were unwilling to make, and have *honoured* them;

—women have shown greater initiative in villages where scarcity of firewood and fodder is acute as it is they who suffer hardship in collecting these from distant places.

Beyond this, however, there has been nothing 'automatic' in the extent of women's active participation in the development of village common lands, no matter how acute their hardship of searching for fuel and fodder. Even in the villages where women took the initiative and played a leadership role, this was preceded by enabling them to interact with other women's groups through *melas,* visits, training programmes and awareness generation camps. Continuous interaction with the PEDO's women staff has been another crucial input for facilitating women's genuine participation.

Where either the opportunities for exposure or inputs by the women staff have been weak or limited, village women's participation in decision-making has also remained weak. The tradition of women being assigned a passive role in decision-making in non-domestic matters and their isolation are just too powerful for women to break through on their own. In most cases, external catalysts have been needed to jolt them out of their passivity and isolation.

But, as evident from women's limited participation in villages such as Wagbole (several other groups—Oopar Gamia, Gundikuan, Wajela—are similar in this respect), periodic inputs and exposures are also not enough. What is needed is to help women create their own forum in the villages where they can meet and interact *regularly* to build up their confidence and develop the capability to work collectively as a group. Initial efforts to create such groups around wasteland development brought forth men's suspicions and hostility as it was a threatening land-based activity. Modifying this approach to work with combined groups of men and women proved a setback to women's participation. The majority of women (there are always a few exceptions) feel too inhibited to participate equally in groups where men are also present.

This much needed women's forum has finally emerged in the

form of Women's Savings Groups. For the first time, women have been able to find a socially accepted space where they can get together *regularly* and have access to information on a *sustained* basis. Given the opportunity to participate in an uninhibited manner in exclusive women's meetings, they are increasingly gaining the confidence to articulate their views even in meetings where men are present. Indirectly, this has increased their genuine participation in the management-related decisions of their common lands. In fact, the pattern which is emerging is that once women get mobilised to act on a particular problem, they leave the men trailing behind in initiative and commitment, sometimes having to drag the men along. This was evident in a few villages when the women decided to take collective action against corrupt LAMPS officials and the exploitative practices of moneylenders. The men were too scared of the repercussions of confronting their exploiters and the women often had to drag their men along as the loans are in the names of men!

Another pattern which is emerging is that women tend to be less manipulative in their actions. Due to being sidelined by the increasingly corrupted maledominated leadership structures and government services, they have not absorbed those values to the same extent. They are more prone to honouring the commitments they make.

Changed Views of the Male Staff

This is illustrated by the changed views of PEDO's male staff about increasing women's participation in PEDO's wasteland development and other programmes. Their initial doubts and hostility of three years ago have changed to their specially going out to involve women simply because women's response is more rewarding. Most of the male staff now feel that women's real, and not just symbolic, participation is increasing and that members of women's savings groups have become more articulate and uninhibited in plantation-related meetings. Many of them are now trying to get more such groups started in their field areas to enable women to gain the necessary confidence. Interestingly, one of them suggested that the tendency of some men to bring the savings money on behalf of their wives should be curbed as men's presence in the women's meetings is inhibiting for the yet weak groups. Once they have crossed a certain threshold, there is no stopping the women. The male staff gave a number of examples of such initiatives by women. In the plantation group meetings in Talaiya village, more women participate now than men. Whereas women take active interest, the men come only to get their presence recorded symbolically. And this is in a village where, to begin with, women were scared to come out and the men questioned the need to involve women. Today, the village men's resistance has diminished as they have witnessed positive changes even within their homes since the women started getting opportunities for going out.

In Behana, one man was trying to misuse the labour musterroll. The women organised a meeting, took over the muster-roll and threw the man out. Women generally do not try to manipulate the organisation for quick, personal gains.

In the Nawalshyam plantation, all the 44 labourers who worked on the land were women. Work was done according to specifications and went smoothly. As the person in charge of the FC said, women work with greater conscientiousness.

In Baletighati, where the women took the initiative for the plantation, they are continuing to fight against manipulative tendencies by some of the men.

Overall, there has been a great change in the attitude of the male staff. Compared to their initial doubts, today they recognise the powerful role women can and have started playing. A similar attitudinal change is becoming evident even among the men of several village which have strong women's groups. The women's programme has once again started toying with the idea of experimenting with wasteland development by setting up exclusive women's groups as a means of enabling the women to acquire rights to some land, albeit as a group. If this happens, it will be interesting to see if there is any difference in the way the women's groups manage their lands compared to the earlier mixed groups.

The Status of the Plantations and their Future Management

Some of the common land plantations supported by PEDO

are now four to five years old. In these, as well as in the more recent ones, protection by the community has been excellent. Many are beginning to resemble fresh, young patches of forest. When one sees the poor condition of similar plantations undertaken by the Forest Department or the *panchayats,* one can recognise the superiority of the approach developed by PEDO.

However, new questions about the future management of these on a sustainable basis are now beginning to arise. The common funds of some plantation groups are either finished or nearly finished. None of the plantations are mature enough to start yielding substantial cash incomes yet. There is also the question of whether they should be used for generating cash incomes at all. Where will the money for paying the *chowkidars* now come from? Will the people be willing to contribute either cash or labour for this? What system will be used for harvesting and equitably distributing their produce (timber, grass, fuelwood, seeds, leaf fodder, etc.) among group members? Will the produce be distributed on payment of cash or free of cost? What will the role of the community and the village committee be in taking decisions on these matters? What are the legal rights of the groups in relation to the *panchayats?*

These and related issues were discussed during three-day plantation camps held with ten of the older village groups between September 1989 and March 1990. These camps helped initiate a process of reflection on some of the long-term management questions related to the common land plantations.

Each of the camps was begun with a symbolic commitment to protect the plantation by the villagers tying rakhis to young plants and worshipping them. Some of the issues which were discussed were:

—Grazing versus stall-feeding: Most villagers made a commitment to switch to stall-feeding so that the cattle do not lose as much energy as they do while grazing and more dung becomes available as organic fertiliser.

—Protection after the group's common fund is finished: The villagers, with women in the forefront in many villages, expressed their commitment to continue protecting the plantations with voluntary labour if necessary.

—Priority to local consumption of the produce or to selling it off: In all the camps, the villagers made a commitment to give priority to use produce from their plantations. Selling anything to outsiders would be considered only where there was no need for a particular produce among the group members.

—Harvesting and distributing systems: Many groups have already evolved a variety of systems for harvesting and distributing grass from the plantation. For firewood, it was thought that in the initial years, whoever does the pruning of the plants should have the right to take away the prunings. Once more firewood becomes available, the groups will get it cut by selected members and distribute it equally among all members.

There were interesting discussions on members' rights to timber. Women of the Bhimsor group did not want to discuss any tree cutting whatsoever from the plantation. They wanted all the trees to be protected. In other groups, it was felt that some timber could be cut in the future for meeting personal requirements of disadvantaged members. In any case, no one would be allowed to sell any timber from the plantation. Whoever was permitted to cut a tree would first have to plant five to 10 new ones.

Legal Rights to the Land

Most of the land which the groups have developed is either village *chernot* land or revenue wasteland. In each case, a distinct group of villagers, usually the residents of a *phala*, claim traditional grazing rights which are recognised by other villagers and the *gram panchayat*. All the groups have got a resolution from the *sarpanch* of the *panchayat* saying that the *panchayat* has no objection to the group's planting on the land and all rights to the subsequent produce will belong to the group and not to the *panchayat*.

However, there has been some concern within PEDO about the groups' legal rights to the produce. At present, no one is particularly interested as there is little of value on the land. But once there is a good stock of valuable trees, what will happen? Will the Forest Department be able to prevent the members from harvesting the produce under some provisions of the Indian Forest Act?

Will the *panchayat* be able to stake a claim to the produce as the land is supposed to be under the *panchayat's* jurisdiction? Most of the *gram panchayats* cover several villages and if this happens, group members who have protected the plantations with so much effort will get a minuscule share of the returns.

This issue was also discussed during the plantation *shibirs*. In all cases, group members were emphatic that they would not allow the *panchayat* or anyone else to take away their plantations from them. But a number of complex legal issues remain unresolved and unclear.

Summary and Conclusions

PEDO's initial experience of involving women with wasteland development in a conservative social context brought forth hostile reactions from male members of the community. In their isolation, the women found it difficult to counter this hostility. The organisation had to compromise and start working with combined groups of men and women. However, women's participation in such combined groups remained limited due to their traditional exclusion from such forums. It was not till the women's self-confidence had been built up with the help of a vast range of other facilitative activities that their real participation started increasing. However, once women crossed a certain threshold of breaking out of their traditional roles, their commitment and involvement are proving far greater than that of men. In some groups, women are effectively in the lead and their actions are beginning to inspire similar dynamism in the women of adjoining villages.

3 *Women Organise for Wasteland Development: A Case Study of SARTHI in Gujarat*

MADHU SARIN
RENU KHANNA

INTRODUCTION

Wastelands development exclusively by women's groups has become one of SARTHI's (Social Action with Rural and Tribal Inhabitants of India) major programmes in Santrampur block of Panchmahals district, Gujarat. The uniqueness and strength of the programme stems from the fact that it originated from the women of Muvasa village themselves articulating the need to find some means of solving their acute problems of fodder and firewood scarcity. This was during the third successive drought year of 1987. Working with women's groups was not a component of SARTHI's wastelands development programme till then. It is to the credit of two field staff members of SARTHI, Champaben[1] and Vikrambhai,

[1] Starting as a shy and apprehensive local woman who had to be coaxed into travelling to Mada (three hours away by road) to attend a *chulha mistris* training programme in 1984, Champaben blossomed into a wonderful, sensitive activist-cum-trainer. She became a key resource person for other women's programmes initiated by SARTHI. Sadly, Champaben died in December 1989 at the young age of barely 30. We, along with other staff members of SARTHI, would like to dedicate this case study to Champaben's memory which continues to inspire all those

that they were able to respond to the situation and lay the foundations of the present programme. As members of a small professional resource group, both the co-authors of this case study have been associated with SARTHI since 1983 and 1985 respectively. Some of SARTHI's staff attended our workshops with PEDO's staff in Mada and we conducted a number of planning and review workshops on the wastelands development programme with SARTHI's staff during 1988 and 1989. It was after two training workshops in Mada in November 1987 and January 1988 that the Muvasa women's group was formed. We learnt about it a few months later. Both workshops were attended by Champaben who played a crucial role in developing the concept of working with women's groups at SARTHI.

As described in the case study, there continue to be several unresolved questions, particularly with respect to securing the groups' legal rights to the land. However, even the limited experience of two and a half years of working with women's groups seems to indicate that the processes of selecting the land as well as group members on the basis of consensus through open discussion with other villagers can be very effective in dealing with the normal resistance to women's involvement with land-based activities. Similarly, enabling women to obtain social approval for taking group action to deal with a problem specifically affecting them has been a tremendously empowering experience for the women. They have been able to create a forum for themselves (non-existent traditionally) to meet regularly where, besides discussing development and management of their common resource, they are beginning to address several other problems affecting their lives.

The Socio–Economic Structure of Panchmahals District

SARTHI is based in Godhar village in Santrampur block of the

---

who were privileged to work with her. We would also like to express our gratitude to Giriraj Singh (Director, SARTHI), Nirmalben (In-charge, Women's Programmes), all those in charge of Field Centres and the field staff of SARTHI for all their cooperation during our field visit for this case study. We would also like to thank members of all the women's groups who gave us much of their valuable time during the peak sowing season and shared their feelings and perceptions with us.

Panchmahals district, Gujarat. It started working there in August 1980 as a sub-centre of SWRC (Social Work and Research Centre), Tilonia. In October 1985, it attained independent status as SARTHI. Its main area of work is Santrampur block although it also has a sub-centre in Savli block of neighbouring Baroda district.

Panchmahals is the second largest district of Gujarat. It is also one of Gujarat's most industrially backward districts with 89.15 per cent of the total workers still dependent on agriculture and allied activities. Surprisingly, there was no change in this percentage between 1961 and 1981. Out of the district's total reporting area of 885,000 hectares, 473,200 hectares (53.5 per cent) is under cultivation. However, only 44,700 hectares (9.45 per cent of the cultivated area) is irrigated. The main source of irrigation continues to be wells. Thus, over 90 per cent of the land under cultivation is rainfed. This leaves the district's economy highly vulnerable to crop failures due to deficient or untimely rains. About 84 per cent of the district's main workers are cultivators compared to about 50 per cent for the state as a whole. Most of them are small and marginal farmers. The district has the lowest percentage of agricultural labourers—only four per cent—compared to 14.8 per cent for the state as a whole.

The average annual rainfall in the district is about 1000 millimetres; 90 per cent of this is received between June and September. However, variation in the annual rainfall is large and six out of every ten years are prone to drought. The District Gazetteer records major famines and droughts during the last and present centuries.

The most recent drought years were 1985 to 1987 which brought acute misery to the people. Besides total crop failures, acute fodder scarcity resulted in heavy cattle mortality and forced large numbers of people to migrate to other areas in search of work. In fact, seasonal migration for wage work is common in the area as rainfed agriculture yields only one crop annually.

Some data about the district compared to the State of Gujarat as a whole according to the 1981 Census are summarised in Table 3.1.

Thus, compared to Gujarat as a whole, Panchmahals has a higher density of population, a lower literacy rate (particularly among women), less urbanisation and a higher percentage of tribal population. The slightly lower population growth rate and the

Table 3.1

Characteristics of Panchmahals District Compared to Gujarat State as a Whole

	Gujarat State	*Panchmahals District*
Total population	34,085,799	2,231,689
Population growth rate (1971–81)	+27.67	+25.58
Density of population (per sq. kilometre)	174	262
Sex Ratio (Females per 1000 males)	942	959
Literacy rate		
Persons	43.7	28.1
Males	54.44	40.79
Females	32.30	14.86
Percentage of urban population	31.10	11.09
Percentage of Scheduled Tribes	14.22	41.76
Percentage of Scheduled Castes	7.15	3.80
Area under forests	9.96 per cent	25.5 per cent

Source: Census of India, 1981.

more favourable sex ratio are possibly due to male dominated migration from the area.

The State of the Forests

The forests of the Panchmahals are historically famous. It was here that the Sultans of Ahmedabad and the Mughal Emperors used to come on elephant-hunting expeditions. Its teak wood was (and is) much in demand. Among all the districts of Gujarat, Panchmahals boasts of the largest area (25.5 per cent) under forests.

Unfortunately, little of the forests has survived the combined pressures of illicit felling, commercial exploitation, clearance for agriculture and uncontrolled grazing. With the continuing pressure of growing human and cattle populations, the remaining forests continue to get further depleted. Excessive grazing smothers natural regeneration and each drought year brings a fresh onslaught from a desperate population in search of fodder, firewood and a source of income. The district's grazing *(gauchar)* land is under similar pressure. Out of a total of about 50,000 hectares of grazing land, at least 10,000 hectares have been encroached upon by a land-hungry population. The rest of it is in a degraded and unproductive state due to the absence of any regulatory or management system.

Santrampur Block

Santrampur block falls in the semi-arid zone and its statistical profile is fairly similar to that of Panchmahals district as a whole. A major difference is that out of its total population of 323,268 as many as 66.4 per cent belong to Scheduled Tribes and 2.8 per cent to Scheduled Castes. Out of the block's total area of 1,36,075 hectares, only 52.8 per cent is cultivated, 9.6 per cent of which is irrigated. Wells are the main source of irrigation. Out of a total of 30,215 landholders, 19,667 (65 per cent) are small and marginal farmers. The block has 380 inhabited villages out of which SARTHI has been working in 110.

About 16 per cent of the block's area is supposed to be under forests although, as elsewhere in the district, the forests are now in a pitiable condition. Vast stretches of bare, rocky hills which are visible were reportedly covered with thick vegetation till just a decade or two ago. The dominant forest species is teak interspersed with several other valuable species such as *dhawra, amla, timru,* and bamboo, *Mahuda* is another local tree, important particularly for the tribal population.

About 8.9 per cent of the block's area is revenue wasteland. Its present status, according to records available with the *tehsildar,* is given in Table 3.2.

Thus, almost two-fifths of the block's revenue wasteland is already under cultivation (authorised or unauthorised) and what remains is degraded hilly or plain non-cultivable land. Even this is under constant pressure of encroachment. There is similar pressure on grazing lands by the land-hungry people.

Table 3.2

Present Land-Use Status of Revenue Wasteland in Santrampur Block

Revenue waste land in the block	*Area (ha)*	*Percentage*
1. Total	12,154	100.00
2. Unauthorised agriculture	3,579	29.45
3. Agricultural land	1,023	8.42
4. Non-cultivable wasteland	3,256	26.79
5. Hilly, non-cultivable	4,254	35.00

Source: Taluka Development Officer, Santrampur

As a consequence, no land is considered 'waste' land by the people. This is a major constraint for a wastelands development programme in the area. Due to the majority of cultivators being small or marginal farmers, a similar constraint governs availability of private wastelands.

The Status of Women and their Relationship with the Natural Resource Base

The status of women in Panchmahals, as elsewhere in the country, is low. Their workload is high. In addition to all the housework (cleaning, fetching water, gathering fuel, livestock care, cooking, child care and care of the sick and the elderly) many women work on daily wages either on the fields of big landowners or on public works. Women of landowning families also perform substantial agricultural tasks. While men do the ploughing and marketing of the produce, women shoulder major responsibilities of hoeing, weeding, irrigating, harvesting and processing the produce. In the traditional division of labour between men and women in a predominantly subsistence economy, 'women's work' of collecting cooking fuel, fetching water and cattle care binds them more intimately with the natural resource base than men. It is they who have to pay the price for environmental degradation by having to walk longer distances in their daily search for fuel, fodder and water. With the tightening of control over forest lands, it is women who have to suffer humiliation by forest staff when caught collecting firewood. As a number of women of the area have said, 'Men just expect to be served cooked food. They aren't bothered about where the firewood comes from. That is *our* problem.'

Desertion, domestic violence, alcoholism among men (despite prohibition in Gujarat) and husbands bringing in second wives to beget sons are common problems of women of the area. On top of that, during drought years, they have to shoulder the responsibility of even men's work if men migrate in search of wages. Many women also have to migrate with the men to labour in more prosperous areas. As a consequence, women's lives are surrounded by a constant web of insecurity which gets reflected in any programme of working with them. This became abundantly evident during the improved *chulha* programme started by SARTHI in 1983. The programme continued to face a high turnover of trained women *chulha mistris* primarily due to the recurring personal crises in their lives.

Also, as elsewhere, women own little land or property. These are almost exclusively owned by men. Whether at her parental home or that of her in-laws, despite labouring hard in the subsistence economy, the woman's economic status is like that of an unpaid domestic labourer. Decision-making concerning land is almost an exclusive preserve of men. This makes wastelands development with women a particularly difficult task.

Evolution of SARTHI's Women's Programme

The process of evolution of SARTHI's women's programme has had a significant impact on the emergence of women's wastelands groups.

The improved *chulha* programme initiated in 1983 was SARTHI's first programme exclusively with women. Selected village women were trained as *chulha mistris* to build improved *chulhas* in and near their villages. They also provided regular follow-up support to the new *chulha* owners. By 1987, a village-based cadre of 30 *chulha mistris* and four supervisors had been built up. The four *chulha* supervisors were SARTHI's first full-time women staff. As each *mistri* was in contact with at least 30 to

50 other women, for the first time SARTHI established a large network of contact with local women. Through regular monthly meetings of *mistris* and supervisors, the organisation's understanding about women's problems in the area started increasing.

By 1987, with the third successive monsoon failure, the hardship of the local people had become acute. Through the *mistris* and supervisors, agonising tales of cattle dying, women's unending search for fuel, fodder, food and water and outbreaks of various illnesses and diseases exacerbated by nutritional deficiencies started pouring in. For the first time, serious thought started being given to developing a more holistic women's programme to address women's issues. In April 1987, a group of village women and the women staff went to Mada to attend the women's *mela* there. They came back enthused and excited about the possibilities of organised action. In November 1987, the four women supervisors attended a training programme for women staff at Mada. Besides sharing their experiences as women, the participants discussed possibilities of women taking collective action to solve some of their problems. And, for the first time, the women participated in an exercise analysing the environmental crisis—its causes and what women could do to reverse the trend.

Although till then SARTHI still did not have a specific programme for involving women in wastelands development, the women supervisors started discussing natural resource management issues with village women during their field visits.

This was further reinforced by the 10-day intensive evaluation-cum-training programme held at Mada in January 1988. Eight staff members of SARTHI participated in this. The process of formation of the first women's wastelands group in Muvasa started soon after this.

In July 1988, SARTHI conducted a training programme for *dais* (midwives, both traditional and modern) in response to the women's demand for access to improved health services. With this, a second cadre of trained village women was created. The *dais* were asked to hold regular meetings in their villages to monitor women's health problems and assist those whom they were unable to help to gain access to the government infrastructure.

Through these village meetings, acute nutritional deficiencies, particularly of Vitamin A, leading to night blindness among children and pregnant women, became evident. This led to discussions on the relationship between health, nutrition and the environment. Integration of nutritionally important trees in SARTHI's wastelands programme started taking place. The trained *dais* have been extremely effective in increasing health awareness among local women.

During 1989, while initiating collection of local knowledge about local trees and shrubs to facilitate species selection for the wastelands programme, it became evident through the *dais* that there was a rich local tradition of using medicinal herbs and plants for dealing with women's health problems. Integrating species of medicinal value in the wastelands programme is likely to become yet another new facet of the programme. The forests of Santrampur block are rich in medicinal herbs and plants.

In May 1990, again in response to a felt need articulated by local women, ten village women were trained to work as para-vets (called *dungar dais*). With them, a third cadre of skilled local women has been created. Within a span of barely two months, the para-vet women have organised getting almost 3,000 cattle vaccinated against the foot and mouth disease. Meetings of women's groups have become the forum for all these skilled women to interact with the others and increase their awareness and knowledge about various problems.

In between, a number of awareness generation and leadership development camps have been organised for selected women at Godhar. Camps on clean drinking water and the environment have been held in several villages. The midwives have been taken to a sister organisation for follow-up visits and many women have been sent to *melas*, workshops and camps organised by other organisations on themes relevant to women's lives. Visits between members of different local groups have also been organised. Each opportunity for a village woman to step outside the village and interact with a large number of other women is in itself a very empowering experience for her. Besides acquiring new knowledge, she is able to shed some of her traditional shyness and inhibitions during each such experience.

Today, SARTHI's women's programme is beginning to look like the organisation's major programme. Due to women's positive response to each awareness-raising exposure, SARTHI seems to be on the verge of switching over from individual beneficiary-oriented projects to programmes geared to organising and mobilising women.

The development of the women's wastelands groups has to be seen against the above background. At least some members of the stronger groups have participated in one or more of the several camps, training sessions or visits organised by SARTHI and almost invariably have been instrumental in helping the others overcome their suspicions or reservations about getting organised.

SARTHI's Initial Work on Private Wastelands

Each year, from 1981 to 1986, SARTHI raised a few thousand saplings of fruit and other plants in village nurseries and distributed them free of cost to several villagers. There was little feedback on what happened to the plants due to the absence of any effective feedback or monitoring system.

In 1986–87, with the availability of drought relief funds to generate badly needed daily wage employment for local people, planting was combined with the development of 'wasteland'. The organisation planted on 20 hectares of common land in its subcentre in Baroda district.

Following the evaluation-cum-training workshop on environmental reconstruction held in Mada in January 1988, SARTHI defined some guidelines for developing private wastelands in Santrampur block. Funds for developing 20 hectares were made available by the National Wasteland Development Board (NWDB) and for another 90 hectares by HIVOS (a Dutch funding agency). Norms for payment for pit and trench digging, digging of boundary trenches, planting, hoeing, weeding, etc., were clearly specified and the staff asked to identify interested small and marginal farmer wasteland owners in their respective field centres. The labour rates were kept a little lower than the minimum rates, the balance being justified as beneficiary contribution. There was limited discussion with landowners about protection after planting and no clear guidelines for their participating in the programme were specified. The general objective was to increase availability of fuelwood and fodder through planting appropriate species.

There was no specific focus on involving women although the women staff were asked to encourage women's participation. One thing SARTHI made clear from the beginning was that it would not permit exclusive planting of eucalyptus although some plants of the species would be supplied.

Review workshops with the staff in August and November 1988, after the planting season, indicated that several, almost bizarre, problems had been experienced while implementing the programme. Some of these were:

—Although landowners agreed to participate in the programme, when earthwork started, many of them preferred working on government drought relief works instead of on their own lands. This was partly because SARTHI's staff demanded digging to specifications while on government works supervision was less rigorous. There was also a wide variation in the wage rates paid by the Forest Department, other government departments and SARTHI. People preferred getting higher wages than being paid less to develop their own land. As a consequence, many of SARTHI's field staff were forced to hire outside labour to work on the landowners' land!

—Although earlier the landowners had specified a wide range of species they wanted to plant, at the time of planting, most people demanded only eucalyptus. One man threatened to sue SARTHI for damaging his land unless it planted eucalyptus on it.

—A major reason for hostile reactions among many landowners was a rumour that the organisation would eventually take away both the land and the trees.

—Many plants were stolen from the nurseries. These included *eucalyptus*, *babul*, *khair*, bamboo and *sewan*. A sack full of stolen bamboo rhizomes was found in a villager's house. On being asked why he had stolen them, he said that if these were planted on the private wasteland nearby, he would have no place left to graze his goats!

—Once the planting season was over, few people took interest in protection or completing the boundary trenches.

It was clear that people were not sharing SARTHI's interest in helping them develop their wastelands into a long-term asset.

They seemed to equate the work with other government drought relief projects with an immediate interest only in earning wages.

It was during these workshops that work on four hectares of common land by a women's group was reported by the staff of the Muvasa field centre. Although some conflict with the residents of an adjoining village, who were used to grazing their cattle on that land, was mentioned, it seemed that fewer problems of the type mentioned above were reported with the group.

The staff were encouraged to analyse why so many problems had been experienced with private landowners and to suggest changes SARTHI should make in its wastelands programme to make it more meaningful for the villagers. It was clear that it made little sense for SARTHI itself to get work done on privately owned land without the owners' participation. If the owners were not interested, the organisation should not get involved.

The staff was also asked to consider the pros and cons of working on common lands with groups rather than with individual landowners. Given the predominance of small and marginal farmers in the area, landowners having any 'wasteland' were likely to be among the better off members of the community. Thus, work on such private lands inherently tended to favour the relatively better off. On the other hand, the totally landless or very small farmer families, who most needed assured fodder and firewood supplies, were automatically excluded from benefiting from the programme.

Although the staff agreed with this basis for giving priority to common lands, most of them felt that few, if any, such lands were available in the area. This is due to the high density of population and unauthorised encroachment. Even so, some basic principles for working with groups on common lands were worked out and the staff asked to explore the feasibility of forming such groups. Two staff members were sent to PEDO to learn from its experience of working on common lands with groups. The basis for working on private lands was also radically revised to avoid the type of problems described above.

The Shift to Working with Women's Groups

It was the response of the first women's plantation group in

Muvasa to a crisis in February 1989 which ended SARTHI's field staff's ambivalence about working with groups on common lands. In fact, without the organisation clearly defining such a policy shift, from that time onwards all the field staff started organising women's wastelands groups.

The incident which inspired this change was as follows. The Muvasa group had harvested its first grass crop from its four hectare plantation with voluntary labour. As the group had not yet decided what to do with the grass, it had all been piled up and left on the plantation itself.

One evening, while passing by, three drunken men from the adjoining village threw an unextinguished *beedi* stub on the grass. All the grass went up in flames and the fire spread in the plantation, damaging many of the young plants carefully nurtured by the women.

Initially, the mishap sent a wave of despondency among the women's group and SARTHI. The plantation was the best out of the previous years's effort. But the women quickly gathered their wits and called a meeting to discuss what to do. Were they going to permit the three men to get away with such an act of irresponsibility? Did their labour have no social value? If they did not punish the men, their future efforts could be similarly nullified. The women decided to demand compensation from the three men. The value of the damage was discussed. It was decided that the men should be made to pay Rs. 1400. As they were likely to bargain, it was decided to ask for double this amount and finally agree on Rs. 1400.

Interesting dynamics took place within the group over this issue. One of the women members was related to the men. She tried to plead that they were too poor to pay such a large amount. The rest of the group told her clearly that their decision was final. If she did not agree, she could leave the group. The woman decided to stay in the group and accept its decision.

The men were summoned and told the group's verdict. Seeing the group's firmness and sensing the entire village's sympathies for it, the men paid the Rs. 1400.

When the Muvasa field centre staff narrated this incident during SARTHI's monthly staff meeting, the rest of the staff felt envious and inspired. The women's group's response was different from the bickering, irresponsibility and manipulations of private landowners. From then onwards, all the field staff started striving to replicate the Muvasa group in their own areas.

Overview of SARTHI's Experience with Women's Groups

At present, SARTHI is working with nine women's wastelands groups in four of its field centres (FCs) in Santrampur block. Details about these are given in Appendix 1.1. A few other groups were attempted but either they did not materialise or they became defunct before they could be consolidated. Disruptive village politics and the staff's inexperience in handling such situations seem to have been the major cause of such failure.

The land being developed in all cases is either *panchayat gauchar* land or revenue wasteland. Subsequent to the spontaneous development of the Muvasa group, initiation of women's groups in other villages has taken different routes. In some cases, SARTHI's field staff have first scouted around to find the necessary wasteland. They have then tried to motivate the local women to develop a group. This has sometimes forced them to start work in villages where SARTHI has not done any work before and the villagers are totally unfamiliar with the organisation.

In other cases, seeing the work of the Muvasa group, some male village leaders have approached SARTHI asking for a similar women's group to be formed in their village, and offering the necessary wasteland.

As may be expected, the experience with the groups has been variable. While some are developing well, others are weak. Some have fallen apart in the early stages of their formation. A complex range of factors influence each group's development. These include: power dynamics within the village, competing interests in the common land, women's status and the extent of their exposure to the outside world, men's attitude towards women taking such initiatives, the skill, experience and leadership qualities of the field staff (particularly of the women staff), and the extent of fuelwood and fodder scarcity in the area.

Case Studies of Selected Women's Groups

In this section, we look at some of the groups in greater detail in terms of the process of group formation, the reactions of other sections of the community (particularly of men) to the emergence of the group, the personal changes and benefits the women have derived through group membership and the role they see their wasteland playing for them in the future.

Muvasa Women's Group

As already mentioned, this was the first women's group to be established in the area and it has played a catalytic role in getting other groups going. Bringing women of newly formed groups to meet the Muvasa group has become a regular feature at SARTHI. These visits have seldom failed to impress new group members and have invariably facilitated new group formation. As the Muvasa group evolved through a unique and organic process, it is examined in greater detail.

Muvasa village

Muvasa is a large village of 550 families. It has six *phalas*. The members of the women's group belong to the two *phalas* of Muvasa and Govalia Mahadev. These two *phalas* have 300 families out of which 100 or so are Patels (high caste Hindus). The average landholding of the Patels is 10 acres. Economically, they are the most well-off. Patel women do not go out to work and most of their fuel and fodder needs are met from their own lands. If necessary, they purchase rather than collect these.

Out of the remaining 200 families of the two *phalas*, the majority are *adivasis*, and the others are from backward classes. Their average landholding is two to three acres due to which both men and women have to supplement agricultural incomes with wage work. Some *adivasi* men work as masons, carpenters or well-diggers. About ten per cent work as agricultural labourers on land belonging to the Patels and 30 to 40 per cent of the *adivasi* women also do wage work either on government public works or on the Patels' land.

Fuel and Fodder scarcity and the condition of common lands

Women of small and marginal farmer families face acute hardship in meeting their fuel and fodder needs. During years of good rains, they get adequate grass from their own lands during the monsoon and a few post-monsoon months. Fodder scarcity is experienced during the summer, particularly from April to June. Several women then have to make a daily walk to a canal four kilometres away to fetch headloads of grass. Each trip takes approximately six hours from 5 a.m. to 11 a. m.

Firewood has to be fetched from the forests about ten kilometres away. Women in groups of four to five make daily trips for one week at a stretch, starting at three in the morning and returning around noon and then take a break for about 10 days before resuming collection.

Scarcity of fuelwood and fodder is becoming more acute with a further dwindling of the village's common land resources. Accurate information about the amount of *gauchar* and revenue wasteland in the village is not readily available. However, much of it has been encroached upon. According to one estimate, there are about 60 hectares of revenue wasteland in Muvasa; 15 hectares of this wasteland were allotted to eight families in 1982–83 for cultivation. They have encroached on another 15 hectares. Another 20 hectares seem to have been similarly encroached upon by other families. The Forest Department had planted on the remaining 10 hectares some years ago but nothing survived. The limited land actually left for grazing is under such intense grazing pressure that it has become totally unproductive and barren. It is four hectares of this land on which the Muvasa group has planted.

Formation of the women's group

In early 1988, SARTHI had initiated the development of private wasteland with a few families in Muvasa. Due to many families not working on their own lands, the SARTHI staff had to hire other labourers to do the earthwork. Most of this labour in Muvasa consisted of *adivasi* women from disadvantaged families.

Coming together to work on a programme for which the organisation's goal was to increase fuel and fodder availability, the women talked about their own hardships due to their scarcity. They regretted their inability to participate in the programme as their own landholdings were too small. Champa Ben and Vikram Bhai, SARTHI's field staff, took a lead from this and started exploring the possibility of the women developing a piece of common land as a group. The women showed immense interest. Besides their long-term interest in securing fodder and firewood supplies, they had an immediate interest in the daily wage employment the work would generate during a drought year. SARTHI was willing to pay for the land preparation and planting work on the same pattern as for private wastelands.

Impromptu meetings on the work-sites resulted in some ground rules being worked out for the group. All those who became members would have to do the land preparation and planting work themselves. The group would have to accept collective responsibility for protecting the plantation. Each member would have a right to an equal share of the produce. All decisions related to the plantation would have to be taken collectively. Only women would be eligible for group membership as it is they who suffer due to fuel and fodder scarcity. All women interested in joining the group would be invited to join.

With the group beginning to come together, the *panchayat* was approached to find out if it could make any land available to the group. The *sarpanch* agreed to give four hectares of what he thought was village *gauchar* land for 10 years. The *panchayat* gave a No-Objection Certificate (NOC) to this effect with a commitment to renew the lease after 10 years and with no claim to any share of the produce from the land. Interestingly, several months later, when the land's record was sought from the *patwari*, it was discovered that the land was actually revenue wasteland and not *gauchar* land. This gives an indication of the prevailing confusion about the status of land in the area.

Some of the more active women then went from house to house enquiring from other non-Patel women whether they were interested in joining the group. A total number of 29 women, all *adivasi*, decided to become members. All of them come from the poorest families in the village, with some owning less than one acre of land. The village has about 30 other families with a similar socio-economic status. They expressed their inability to join the group either because of inadequate labour in the family, too much other work or other reasons. Due to this process, no conflict was generated due to some women feeling excluded from the group.

Dealing with men's suspicions and initial hostility

Even before the group started work on the land in April 1988, men of different castes and classes had started questioning why the organisation was working only with women. The husbands of some members objected to their attending group meetings. They felt that housework was suffering and the time spent in meetings was wasted. It seems that at the family level, each woman had to use a combination of assertiveness and diplomacy. As some of them said during a recent meeting, they started getting up earlier to finish their housework before going to work on the land or to attend a meeting. One woman said that her mother-in-law still asks her every time why she is going for a meeting and she has simply learnt to say loudly and assertively that she has to go because her sisters have called her. The sisters-in-law of some women also resent their spending so much time on group work as they have to do additional housework.

SARTHI's staff dealt with the men's hostility by calling a meeting of the whole village to explain why they were working with a women's group. Women's increasing hardships due to scarcity of fuel and fodder were discussed and women having a greater stake in dealing with the problem was explained as the main reason for working with only women. Although this helped reduce the members' husbands' suspicions, other men continued spreading rumours that SARTHI would abduct the women, take away their jewellery, etc. Some men would stealthily listen to the group's discussions.

Slowly, with the group working steadily, the rumours and suspicions died down. SARTHI was also distributing cattle and deepening wells as drought relief work at that time. It had also organised environmental awareness camps. These activities had increased SARTHI's credibility and helped reduce the resistance of the menfolk. The field staff also ensured some men's involvement with SARTHI's other activities so that they were better acquainted with the organisation.

Evolving a management system for the plantation

The group meets regularly on the tenth of each month. Many

decisions about the plantation have been taken during these meetings. The first problem the group faced was getting the co-operation of the adjoining village of Sadavra in protecting their plantation. Earlier, some families of this village used to graze their cattle on the land. This problem was solved with the help of Muvasa's *panchayat* and by persuading the concerned families.

Next, when planting on the land was started in 1988, some village men had spearheaded a demand that only *eucalyptus* should be planted. They tried to pressurise the women's group to make a similar demand. On being reminded about their primary need for fuel and fodder, the women refused to succumb to this. A diverse range of species, earlier selected by the women themselves, were planted.

How the group dealt with the crisis of their first harvest of grass getting burnt in early 1989 has already been described. In November 1989, with the second grass crop ready for harvesting, the women were asked how they were planning to harvest and distribute it. Surprisingly, most of them felt that it should be cut by voluntary labour and sold. Part of the money could be distributed among the members and the rest deposited in their common account. This led to an interesting discussion on the contradiction in this decision. How was it that, although the group had raised the plantation for meeting their fodder needs, when fodder was becoming available from it, they were planning to sell it? The women's initial response was that all of them had enough fodder from their own lands. Then why had they raised the plantation? Was their own fodder going to be adequate for the whole year? To this the response was that while some members would have enough for the whole year, others would run out by the summer. What would the latter do then? Buy from the market at a higher price than the price the group would get at this time? Or would they again walk four kilometres to the canal daily to fetch head-loads of grass? Of what use was the plantation to such members?

The group had not given serious thought to these questions till then but they understood their pertinence quickly. It turned out that because the fodder scarcity was going to be experienced a few months later, planning for it would imply storing the grass till then. Grass being bulky, and one crop of it already burnt to ashes, the women were reluctant to consider storing it on the plantation from where it could also be stolen. Their private storage space was already full with grass from their own lands. Building additional space would be expensive. Lastly, there was the tricky question of different members requiring different amounts of grass due to variation in both landholdings and numbers of cattle owned. In the fear of dissension cropping up within the group over some members getting more grass than others, the consensus was to dispose of it all through sale! The group was asked to decide on the matter after giving it further thought.

The group decided on a compromise solution. Members were invited to cut the grass on the basis of taking away half the amount they cut and leaving the other half for the common pool. Fifteen women cut the grass on this basis. As *dhaman* (fodder grass seeds of a superior quality) sown the previous year had taken root, the women took away this grass as their share. They fed it to their cattle instead of storing it and felt it had improved milk yields. The poorer quality local *hukra* grass, which is very prickly, was left for the common pool. This was sold to a villager at Re. 1 per four bushels (the prevailing market rate was Re. 1 per three bushels but the group's bushels were smaller and the grass of poor quality). The group earned Rs. 250 from the grass. About ten kilograms of *dhaman* grass seeds were also collected from the plantation.

At a meeting in July 1991, the group was asked how many members had to buy grass this summer. Sheepishly, they admitted that every single member had to buy at least some grass and that too at the high rate of Rs. 3 a bushel! They said they had understood the meaning of the questions raised the previous year and that this year, the group would not sell any of its grass.

When asked similar questions about firewood, the women firmly assert that when firewood starts becoming available, it will *definitely not* be sold. They face firewood shortage scarcity all the year round and fetching it is even more arduous than fetching grass.

During 1988 and 1989, the group had experimented with dibbling some seeds of beans like *arhar*, *urad* and *chaula* in the plantation. Being nitrogen fixing, they improve the soil besides covering bare soil quickly during the monsoons. In addition, it was hoped some nutritious edible beans would be produced.

The women monitored the experiment with interest. There was some survival of *tuar*. About two kilograms of seed were planted and 15 kilograms harvested. Besides, many women plucked green beans for cooking. However, *urad* and *chaula* were eaten by rats and rabbits. This year, the women are planning experimental planting of a local fibre yielding plant called *bhendi*. They expect this to perform better than the beans and also yield some quick monetary returns to the group.

The women also have a lot of knowledge about local trees and shrubs and their suitability for different uses. After compiling this information systematically, SARTHI is planning to use it as a basis for species selection by different groups in the future. SARTHI has also started a Seed Bank for collecting seeds of useful local species for its nurseries. Some members of Muvasa's group have been earning some income by collecting seeds for the bank.

Protection of the plantation continues to pose a problem during the fodder-scarce months. As the land is at some distance from where the women live, they cannot easily keep an eye on it while doing other work. The protection trench and the thorny plant fencing is not adequately effective yet. Most of the year, other villagers respect the group's efforts and keep their cattle away. But, it seems, some people cannot resist the temptation of letting in their cattle during the dry summer months. This year, the group hired a watchman during the summer months and is planning to do the same the following year.

After the training conducted by VIKSAT (Vikram Sarabhai Centre for Science and Technology), it was decided to get the group registered as a Tree Growers' Cooperative and then apply to the District Collector to allot the land to the cooperative. All except three of the members have already deposited their share money. SARTHI submitted the group's application for registration in August 1989. The cooperative has still not been registered.

In 1989, encouraged by their achievements with the first patch of land, the group approached the *sarpanch* to give them the remaining six hectares of vacant revenue wasteland for additional planting. Although the *sarpanch* gave them a No-Objection Certificate (NOC) on the basis of which the group started working on the land, this time getting the land did not prove so simple. This land was being used for grazing by four to six families who had earlier got some revenue land allotted to them and have encroached on some more. These families fear that if the group plants on the land, they might eventually be forced to give up the land they have encroached upon. They pressurised the *sarpanch* to withdraw the NOC. On learning this, the women tried fighting back by calling a meeting of the whole village and asking why they were not being allowed to plant on government-owned land. Work has had to be suspended for the time being. The women are hopeful that after registration of their cooperative, they will be able to get the land allotted to it and recommence the work.

Besides discussing plantation-related matters, the group's meetings have become an effective forum for discussing other problems of the women members and increasing their access to information and knowledge. One member is a *chulha mistri*, another a trained midwife and a third one was recently trained as a para-vet. Each of them provides guidance to the group in their respective fields.

Several members were provided buffaloes by SARTHI as a drought relief measure. These seem to have changed the women's lives significantly and are providing multiple benefits. Now the women get good organic manure besides having dung as an alternative fuel when they are unable to fetch firewood. Their agricultural production has increased due to the manure and they are also able to sell milk to the village dairy twice a day.

Income and employment benefits from the plantation

Work generated by a small plantation on only four hectares of land has made a surprisingly significant qualitative impact on the 29 group members' income and employment options. This has largely been due to three rules evolved jointly by SARTHI and the group:

1. That work on the plantation (although paid) will only be done by members of the group or, at the most, by members of their families. In Muvasa's case, almost all the work has been done by the women themselves.
2. Payment will be made on the basis of output (number of pits or trenches dug, plants planted, etc.) and not on the basis of daily wages. The group will monitor each member's work output itself.
3. Flexible work hours will be maintained. Women can vary the amount of work done in a day and their working hours according to their convenience.

Work on the plantation, although diminishing each year, has

been spread over three years so far. During the first year, earthwork and planting were done. During the second year, some additional planting and mortality replacement were done. And this year, the third one, strengthening of the fencing, mortality replacement, etc., are being done. If the group is able to get an additional six hectares of land allotted to them, then work on this flexible pattern will be generated for another three years.

It needs to be noted that the work generated is only seasonal. However, earthwork can be done during the lean agricultural summer season. That is when small and marginal farmers dependent on rainfed agriculture are most in need of wages.

The women described the following changes in their lives due to the above. Before the plantation was started, many of them used to work on daily wages either on the Patels' land or on government works. This was done primarily during the lean summer months. On government works, they seldom got the minimum wages (due to corruption) and work sites were often at considerable distance from the village (many of the women had worked ten to twelve years ago on the Kadana dam several kilometres away). The Patels demand work from sunrise to sunset (at least 8 a.m. to 6 p.m.). Three years ago, they were paying Rs. 5 per day plus two meals and two teas.

The problem for the women was that if they went for wage work on the above terms, they could not do any housework at all. Only husband *or* wife could go out to work as one of them had to stay behind to take care of the children, cattle and other domestic chores.

Ever since the plantation work started in 1988, *none* of the group members have worked either for the Patels or on public works. Now, they can go for wage work late in the morning after finishing their household work. They can easily earn Rs. 10 to Rs. 11 per day. On days they are able to put in more labour, they can earn much more.

According to the women, the Patels are suffering from labour shortage as a consequence and some of their land now remains uncultivated.

Another benefit of flexible work hours is that the women can go to the dairy to sell milk twice a day. The dairy's milk collection timings are fixed and conflict with full-time daily wage work.

Women's perceptions of change in themselves

Getting organised as a group and taking collective action has clearly been an empowering experience for the women. They feel that they have established their right to access to information and knowledge which was earlier available only to men. Collective action enabled them to break through men's initial hostility to women treading into a non-traditional domain. Individual members could deal with such problems at their individual levels knowing that they had group support behind them.

The women recall that initially the husbands of some women would come and forcibly pull their wives away. Group condemnation of such humiliation of its members slowly reduced such behaviour among the men.

The group's strong stand over fining the three men who had carelessly started a fire in the plantation was generally supported and almost admired by the village men. Today, the husbands of many members genuinely feel that the women should get more opportunities for acquiring new skills and knowledge. This is partly due to the change for the better within the members' homes since they joined the group. Some men are still not fully convinced about the desirability of women taking such initiatives. However, they do not openly object in the hope of getting personal benefits through SARTHI in the future.

The group clearly enjoys the leadership role it is playing for members of new groups brought to interact with them. Despite their busy routines and heavy workload, group members always make time for those visiting them. And their newly found eagerness to learn and share is unfailingly inspiring.

Sadavra Women's Group

Sadavra is next to Muvasa and the three men who accidentally caused a fire in the Muvasa group's plantation belong to this village. Because of the proximity between the two villages, Sadavra residents have watched the development of the Muvasa group from close quarters. This is a relatively small village of about 60, largely *adivasi* families, most of whom are small and marginal farmers.

Interestingly, the initiative to form a women's wastelands group in this village was taken by four male leaders. They approached

SARTHI's FC In-charge to form a women's group similar to Muvasa in their village. One of them had already identified four to six hectares of the village's *gauchar* land for the plantation. In retrospect, it is clear that at least some of these men were more interested in personal gain than in having a strong women's group in the village. One of them, Manibhai, who was taking the maximum initiative, has already boycotted the group and SARTHI because he was not given the work of raising the nursery for the plantation. The group identified two of its members as more suitable candidates for raising nurseries! This tendency of men trying to use women's groups for personal gain will need to be carefully countered by SARTHI.

It was Manibhai who first discussed the idea with the village women. The women had already started thinking along the same lines after observing the Muvasa group. They agreed readily.

Many village men also supported the idea probably due to the immediate attraction of daily wage employment being generated in the village. Others, however, are suspicious of SARTHI and unwilling to let their women go out for camps or training programmes.

To counter the men's suspicions, a joint meeting of men and women was called to explain the objectives of the programme and why a group of only women was being formed. It was clarified that all the produce would belong to the group and not to SARTHI. The villagers were also informed about other SARTHI or government programmes such as poultry farming, animal husbandry, etc., which could benefit the villagers.

Initially, 22 women agreed to join the group. They live closest to the land and are from the most disadvantaged families. The husbands of two of them obtained an NOC from the *panchayat* and *patwari* for the women's group to develop the land. The NOC does not specify any time limit for which the land is allotted to the group and does not demand any share of the produce for the *panchayat*.

Sixteen of the 22 members did the land preparation and earthwork. Six women have neither worked on the land nor come for the meetings due to having too much other work. When asked whether they want to remain in the group, they continue to say 'yes. Eight to ten non-member labourers also did some work on the land. The 16 regular members have been attending all group meetings and decided on the species they want to plant. These are mostly fuel and fodder species. The women are confident that they will be able to protect the plantation and have agreed to put aside a small percentage of their wage money for starting a group fund.

All the women earlier used to work on daily wages for the Patels or on public works. This year, they only worked on the plantation. They said that they could earn Rs. 20 to Rs. 25 per day working on their plantation. The Patels pay them only Rs. 5 plus meals but demand at least ten hours of work.

The group is still in the process of consolidating itself and the future of its plantation will depend on the extent to which it is able to handle a number of unresolved conflicts in the village. Some of these are:

1. A number of men still view the plantation as a means of earning easy wages without any accountability for work output. A husband and wife team of non-members was found to be digging pits and trenches of a much smaller size than required. When the FC In-charge pointed this out to group members, they asked the two to either do the work properly or stop working. The man was furious and left the work. His wife is interested in joining the group but the man has started talking against SARTHI.

When we met the group, this woman had also come to attend the meeting. Her husband came in search of her and shouting loudly against SARTHI, pulled her away by force. The group watched the whole affair in embarrassed silence but did nothing. When asked what they had felt on seeing another woman humiliated like that, all the women said that they had felt very bad. The man was angry because his mistake had been pointed out. But didn't the women feel that they should have supported the wife in her desire to attend the meeting? They all agreed that they should have done so. However, it is likely to take them some time to develop the strength to provide group support to individual women. Angry men like this one and Mani bhai who did not get the nursery can be very destructive for a women's group before it has found its feet.

2. It seems that the *gauchar* land the group is developing is the only grazing land of the village. Although the women claim that only cattle of families living near the land (and most of them are members of the group) grazed on the land, it is likely that

during the dry summer months, other families also leave their cattle there. The women members have already started stall-feeding their cattle and even purchased fodder to do so. They are confident that their cattle will not graze on the land. But will summer scarcity bring out the resentment of other families over their only grazing land being closed? Only time will tell how this underlying conflict is resolved.

3. Some men from families of non-members continue to be suspicious of SARTHI and are spreading rumours that SARTHI will take away both the land and the trees. Among them are one or two who had an eye on the land. One of them had hired a tractor and ploughed a part of the land the group is planting on.

The Muvasa group faced similar problems during its initial stages but managed to come through with strength. How the Sadavra group is able to deal with them will determine its growth. The staff's skills in helping minimise conflict will play an important role. The absence of Champaben's leadership skills is being strongly felt in this FC.

Wankadi Women's Group

This group was formed early this year and has 26 members. The group is developing ten hectares of village *gauchar* land for which they have obtained **an** NOC from the *panchayat* and *patwari*. Wankadi falls in a different FC from the earlier two groups. Although barely six months old, the group has attained a strong identity and many of its members display good leadership qualities.

Wankadi is a fairly large village with about 300 households. All of them are either *adivasis* or belong to backward classes (*Buxi Panch* in Gujarat). The village has a number of fairly independent *phalas*. Group members belong to two *phalas* which are quite close to each other.

The majority of families are small or marginal farmers practising rainfed agriculture. Some families have irrigation wells. Members of many families have to migrate during summer months in search of work.

The village is surrounded by hills which are totally denuded. As a consequence, there is an acute scarcity of firewood and fodder. This group's primary motivation for developing the *gauchar* land is to reduce the members' hardship in scrounging for firewood and fodder.

Interestingly, two years ago, SARTHI had attempted organising its first women's group in another *phala* of Wankadi. Men were allowed to attend the women's meetings. The venture was a total disaster. A male leader tried to exploit the organisation by insisting that it distribute goats to the villagers. When SARTHI refused, he instigated the yet to be consolidated group to stop cooperating with SARTHI.

Learning from that experience, this time, after the initial combined meeting with both men and women in which the reasons for forming a women's group were explained, the staff was careful in ensuring that only women attended subsequent meetings. This was accompanied by organising a number of educational inputs for group members in quick succession:

—All the 26 members were taken to meet the Muvasa group. This proved very inspiring and helped the women to define rules for their own group's cohesive functioning.
—Two members of the group attended the training conducted by VIKSAT. They later took an active part in getting the *panchayat's* NOC for the land and explaining the importance of group registration to other members.
—Another member attended a Women's Awareness Camp held at Godhar and a meeting of members of Women's Savings Groups held at Mada in May 1990.
—One member has been trained as a para-vet and has started treatment of common cattle ailments.

The growing strength and coherence of the group is evident from the following decisions and initiatives it has taken on its own:

—The women did most of the running around for getting the NOC from the *panchayat*. SARTHI's staff only provided some guidance.
—The group has decided to get itself registered as a cooperative. Each member will get her share money deducted from the

wages due to her. The women have also decided to put aside a small share of their wages to start a group fund to pay for protection later on.

—Only group members themselves and some members from their families worked on the land. Later the group found that the men (even from their own families) were not doing satisfactory earthwork. The women told the men to stop working on their land! Now all the work is being done by the women themselves.

—Six of the group's members have migrated to an adjoining district due to economic hardship. People from the village go to harvest the *bajri* crop there and get a share of the harvest as their wages. Before going, these six women discussed their compulsion for migrating for two to three months with the rest of the group and said they would pay their share of the group's savings on their return. The rest of the group has agreed to let them rejoin as members on their return.

—During our meeting with the group, an elderly man came and sat down at the other end of the room to listen to the proceedings. One of the group members firmly requested him to leave as in men's presence women have to cover their faces because of which they cannot participate in the discussions properly. The elder and three other men who were also listening hiding behind a *charpoy* meekly got up and left the room!

—The group has won the support of men of their families and of the *panchayat*.

—They are very clear about the need to protect their plantation. Due to complete destruction of forests in the area, women now have to burn *tuar* stalks as fuel or scrounge around the hills to dig out roots.

—The group has itself selected the species to be planted. These are mainly for firewood and tree leaf fodder. They do not expect good grass as the land is poor and stony.

—The group has decided that a fine of Rs. 25 will be imposed on anyone whose cattle gets into the plantation.

Gamri Women's Group

This group was also started early in 1991 at the villagers'

request. They had heard about the Muvasa group. Gamri is a predominantly *adivasi* village in a relatively more interior area. As a consequence, there are few government services and the villagers are less exposed to the outside world.

Twenty-four women from two *phalas* of the village have decided to develop four hectares of village *gauchar* land. The Forest Department had planted on this land some years ago but little survived. The women are determined to protect their new plantation.

Soon after the group was formed, SARTHI's male FC In-charge found that only men from the group members' families were coming to work on the land. The men kept insisting that the group was a women's group but no women were coming for work! During our recent meeting with the women, it turned out that due to the stony nature of the land, the women had found digging very tough. They requested their men to do the earthwork on their behalf. Once planting started, the women had started working on the plantation again.

The women seem very keen and highly motivated to develop firewood and fodder resources near their homes. At present they have to walk five to six kilometres to the nearest forest for fetching firewood. However, it seems that the men are a bit wary if only a male staff member comes to conduct meetings with the women. More men than women show up for these meetings. This has been happening due to Champaben's untimely death. The new woman staff member is not very experienced or confident yet. Ensuring the presence of a woman staff member during future meetings will be important for building up the group.

Relatively Weaker Groups

Most of the weaker groups have some male village leaders interfering with the women's group formation process and relatively inexperienced staff unable to deal with the situation. In some cases, relatively fresh staff with limited understanding about group development have taken inappropriate decisions.

The combined group of Kenpur-Wandariya falls in this category. Although the group was formed a year and a half ago and planted on their village *gauchar* land in 1989, the group continues to be weak. At the time of the group's formation, the *sarpanch* had insisted that some of his women relatives from Kenpur must be made members of the group. The land is closer to Wandariya. At that time, this being the first group in this FC, the staff agreed to this arrangement. Unfortunately, it has become a built-in constraint for the group. The women find it difficult to meet regularly as the villagers are far apart. If the meeting is fixed in Kenpur, many members from Wandariya don't show up and vice versa.

Surprisingly, there has been no problem with the plantation's protection despite the group's weakness.

Another new group which is likely to have problems is that being formed in Sarar village. It seems that from the beginning, the *sarpanch* has insisted on playing a dominant role and does not really let the women have their meetings freely. An inexperienced staff member got the plantation work started without the group really being formed.

Two cases of failure

It is also useful to mention two cases of groups which failed. These failures have provided useful lessons for the future.

A group in Ladpur village was formed in 1989 and undertook planting the same year. Unfortunately, during a crucial phase of group formation, the FC's woman staff member was hospitalised. The male FC In-charge possibly relied too much on a powerful local male leader to help form the group.

Once the plantation was done, this man demanded that SARTHI instal a hand pump on his private land. As this is against SARTHI's policy, he was refused. The man used his influence to prevent the group from cooperating with SARTHI. Protection of the plantation has become a problem and the group has become defunct.

In the second case of failure in Vajyakoot village, an inexperienced young staff member, who had just taken over, selected land in several pieces for planting by a group. Protection of several land patches would have proved difficult in any case. But higher caste Patel families were also in unauthorised occupation of this land. The *panchayat's* NOC was obtained by the staff on their own without first forming the group. When work was started, the more powerful Patels obstructed it. Both the plantation and the group had to be abandoned.

A Synthesis of SARTHI's Experience with Women's Groups to date

As evident from the above review of SARTHI's two and a half years experience with women's wastelands groups, although it has been somewhat variable, sometimes lacking consistency, it indicates tremendous potential. Its strength lies in organising and mobilising women to take positive action to deal with one of their acutest problems, that of firewood and fodder scarcity. The responses of the stronger groups have to be seen in a context of continuing degradation of surviving forests and several government wastelands development programmes which yield few tangible results. The land on which most of the women's groups have planted had already been planted upon earlier either by the Forest Department or the *panchayat*. It needed replanting because the earlier plantations had not survived. Although it is probably too early to assume that the women's plantations will do any better than the government ones, there are strong indications that they will. This is because SARTHI's approach has not been restricted to simply getting a number of plants planted but working with a long-term perspective and helping the groups develop sustainable management systems for the plantations. This has had to be facilitated with a wide range of additional inputs to ensure women's genuine participation at every stage of the process.

The interest of rural women in the subsistence economy in sustainable management of forests and wastelands is clear. However, what is equally clear is that this does not automatically result in women taking initiative to reverse the process of degradation. Traditionally, they are kept too isolated and devoid of access to information to be able to do so. There is also no cultural tradition of women getting together to solve their problems. And then, decisions related to land are considered to fall in man's domain. Breaking through all these barriers to successfully mobilise women is not easy.

Important components of the process evolved by SARTHI to

overcome the normal obstacles to working with women include the following:

1. Seeking social consensus to work exclusively with those women most acutely affected by environmental degradation. Joint meetings of men and women are called in the initial stages to explain the objectives of the programme and seek everyone's support for it. While this ensures that the participating women have a genuine long-term interest in rehabilitating wasteland, it also helps reduce male suspicion and hostility towards work being done only with women.

2. Focusing on developing common instead of private land. Through this, the suspicions of individual landowners are avoided by enabling the women to gain access to a common land resource which is less threatening. Working together, the women get an opportunity to break their isolation and develop collective strength. Taking decisions through group discussions helps build up group identity and accountability to the group. Efforts are made to dissuade men from attending or interfering in the women's meetings.

3. Eligibility for group membership is left open for all local women. All those interested in joining are welcome to do so. As it is made clear early on that membership would imply responsibility for working on the land and protecting and managing the plantation subsequently, women living far away from the land or not having the time to work on the land tend to stay out of the group. Patel and other higher caste women, who have too much work on their own lands and who are socially not permitted to do manual work for wages, also do not join the group. Through this non-conflict generating self-selection process, the women who tend to join the group are those from the poorest families owning the least land, interested in wage work and who face acute hardship in collecting fuel and fodder.

4. Seeking the women's long-term commitment to sustainable management of the resource *prior* to starting any physical work. This involves their participation in planning what to plant (species selection) and evolving a management system for protection and distribution of the produce (equal rights linked to equal responsibilities and group membership being made conditional

to each member conforming to these). The primary focus of the groups' activities has been to undertake plantation activities for meeting the women's subsistence needs of fuel and fodder and *not* generating commercial returns.

5. Facilitating building up the women's confidence through organising visits to other groups, sponsoring participation of some women for awareness generation and leadership development camps or other training programmes that impart skills of immediate relevance to their day-to-day problems. Maximising opportunities for women to go out of their villages and interacting with other rural women who have similarly broken through traditional barriers has been an extremely effective and empowering experience for the women.

6. Converting regular group meetings into a forum for discussing other problems affecting women and facilitating their participation in decision-making in all relevant domains.

Differences between working with women and men

SARTHI's shift from working on individually owned private wastelands to common lands has also involved a shift from working with men to women. We explored the difference SARTHI's staff has felt in working with the two.

Both the male and female staff feel that there is a qualitative difference in working with members of the two sexes. While working on private lands, it was extremely difficult to involve women in any part of decision-making. The women usually wanted to consult their husbands first. Even if they took any decisions on their own, they were often overruled by the men.

The men, due to their greater involvement with non-domestic affairs, are more integrated in the growing culture of manipulation, corruption and bribery. They are more familiar with misuse of government funds and schemes for personal gain by individuals in positions of power (be they in the bureaucracy, local political institutions or local leaders). And most of them attempt doing the same whenever they get the opportunity. Thus, while doing wage work on public works, the effort is to get away with doing the least work possible. Pits and trenches are not dug to specifications, plants are dropped while carrying them from the nursery to get paid for more plants than those actually planted and there is no commitment to long-term goals. There is a strong tendency to make the maximum money *now*. The future will take care of itself.

Similarly, the men here are definitely more interested in commercial species than those which will meet subsistence needs of fuel and fodder. In many respects, the staff's experience of working on private wastelands had some nightmarish aspects. Perhaps confusing SARTHI with the government the men tried bullying the staff to let them have their way, sometimes even threatening them physically. Being asked to dig to the required specifications brought responses of anger and resentment. The women staff were scared to go alone in some areas due to fear of being assaulted by drunken men. They could not rely on any support from local women as they were barely involved with the programme.

The fact that SARTHI's work on private wastelands was primarily concerned with individuals with no accountability to a group also contributed to such a situation.

In contrast, working with women's groups has been totally different. Once a minimal group identity is created and the women have the opportunity to sit together and take decisions, they tend to stick to them. Commonly agreed upon work norms are adhered to and accountability of individuals to the group is high. There is a consistency in decision-making. If they have decided to plant fuel and fodder species to reduce *their* problems, they cannot be swayed into demanding commercial species at the last moment. And given the opportunity to develop their self-confidence, they can throw out even their husbands who do not do the work to specifications (e.g., the Wankadi group).

Beginning with the qualitatively different performance of the Muvasa Women's Group with visible achievements has also substantially reduced male suspicions about SARTHI's motives for working with women. Whereas earlier, men used to pull away or even beat their women for wanting to participate in any of the organisation's activities, now they have started inviting SARTHI to come and start a women's group in their villages. With this, a new space has been created for women to take action and initiatives in fields earlier forbidden to them. SARTHI's staff have also learnt to deal with initial male suspicions and hostility and once this is overcome, they try and support the women to demand that only women attend the women's group meetings. This is to avoid men playing a disruptive role by putting forward unreasonable demands or trying to manipulate the organisation for personal gain. In any case, women continue to feel inhibited in men's presence and as pointed out by a member of the Wankadi group, they have to observe *purdah* in men's presence which limits their participation. In many villages, the situation has changed to the extent that the men encourage their women to participate in other activities offering to do the housework in their absence.

A few amusing comments from men came to our notice during our field visit. These indicate the extent to which women's group meetings have started attaining legitimacy and respectability in the villages. While we were having a meeting with the Padri women's group, the men sat with the male staff at some distance. Apparently, they decided to maintain a safe distance from the women's meeting lest the women asked them to fetch drinking water for the group while its meeting was going on!

In another village, the men were overheard saying that it was best not to try and interfere in the women's meetings lest the women start demanding a right to attend men's meetings!

The male staff also mentioned that women *feel* much worse if a plant dies as it is they who do most of the caring for the plants.

Some unresolved problems related to land

The biggest unresolved problem for the groups is related to land. First, there is an acute scarcity of common lands available for development. The small patches the existing groups have got will be able to meet only a fraction of their fuel and fodder needs, even when fully developed. Second, closing even these small patches to grazing will further reduce availability of grazing lands. Although a switch-over from uncontrolled grazing to stall-feeding is inevitable under changing circumstances, the total absence of grazing land is likely to affect some families adversely. This could become a source of conflict in the future.

But the most difficult problem related to land is its ambiguous legal status. As mentioned earlier, even the village *panchayat* of Muvasa thought a piece of revenue wasteland to be village *gauchar* land till the *patwari's* records were checked.

For the women's groups, however, the matter of greatest concern is how to legally secure their rights to the produce of the land they are developing with so much effort. What is the legal validity of the NOC signed by the *panchayat* and the *patwari*? Can a new *panchayat* overrule the NOC given by the earlier one? Can *gauchar* or revenue wasteland be allotted to a women's group? If yes, what formalities must it complete for getting the allotment? Must it be registered as a cooperative, a society or a *mahila mandal*? Will it get the land on lease? If so, for how long? It is extremely difficult to find helpful answers to these questions. No one seems to know.

A training programme by VIKSAT was organised by SARTHI to throw light on some of these matters. VIKSAT recommended that the groups should be registered as Women's Tree Growers' Cooperatives and then an application be made to the Collector to allot the land to the groups.

SARTHI prepared the papers for the cooperatives on the lines recommended by VIKSAT and submitted applications for registration on behalf of four groups to the Registrar of Cooperative Societies. Even after six months, the cooperatives have not been registered. Each time, there is a new query or a demand for another document.

Recently, we met both the *tehsildar* and the Collector to seek clarification on the matter. Both of them said that the state government discourages the allotment of *gauchar* land for plantation purposes as there is already a scarcity of grazing lands in the villages. However, hearing about the increased grass production due to protection from the lands developed by the women's groups, the Collector offered to convert it into revenue wasteland and then allot it to the groups. But he would like the lease application to be only for five years or to a maximum of 10 years. That is too short a period for such a programme. Firewood and leaf fodder from the trees will barely be becoming available by that time.

As far as revenue wasteland is concerned, surprisingly, the *tehsildar* was himself unaware of the procedure to be followed. He said his office had not received any application for allotment by a group.

The Collector, however, said that the state government had imposed a ban on allotment of any revenue wasteland in the district as this had been reserved for resettling people displaced by the Narmada dam. Any application for allotment of such land would first have to be sent to the Narmada Resettlement Department who would first check whether the land might be suitable for resettlement of the 'oustees'.[2] All the land the groups have developed is stony, hilly, barren and in small patches in the middle of villages. It is unlikely to be considered suitable for resettling dam 'oustees'. But the procedure might take several months prolonging the uncertainty about the legal rights of the groups.

The Collector also recommended that the words 'Tree Growers' be avoided in the names of the cooperative societies. Instead, greater emphasis should be given to fodder production which would make it easier for him to push for allotment of at least the *gauchar* land to the groups.

Whatever the legal complexities, one thing seems clear. Due to the process by which the groups are being formed, especially given the social approval of the entire village for their long-term goals, it will not be easy for any special interest groups within the village to displace them from their common lands. The stronger groups will fight against any such attempt. Securing the groups' legal rights is important, however, from the point of view of protecting them against stronger, outside powers such as the Narmada Rehabilitation Authority or provisions under the Indian Forest Act.

A more promising approach for the future seems to be to organise women's groups to rehabilitate forest lands. The Gujarat Forest Department has recently formulated a scheme under which villagers' groups protecting forest lands will be given rights to collecting all minor forest produce (MFP) and also 20 per cent of the income from sale of timber from the land.

Some Concluding Observations

One of the basic issues concerning women's involvement in wastelands development relates to their resourcelessness in terms of land ownership. Second, their secondary social status denies them opportunities to act equally in non-domestic domains.

SARTHI's approach of organising women's groups for developing common lands does not directly address the issue of women's

[2] 'Oustees' is the official term used in India for persons displaced by public works such as dams, parks, canals, etc.

land-related resourcelessness. If the husband of a group member decides to throw her out, she cannot take her share of the common land with her. However, women's *empowerment* through working in groups is *indirectly* enabling them to gain the strength to fight against such injustices within the family. Not only that, they are gaining the knowledge and confidence to be more assertive in how the family's resources are managed. The wastelands groups are not just successfully developing wastelands but helping in creating a new social space for women which has traditionally been denied to them.

The *process* by which the groups are developed is of crucial importance. Staff skills play an important role in initiating the process. At present, that seems to be SARTHI's main weakness. It must work out mechanisms for imparting these skills to new staff members when the experienced ones have to leave for some reason.

APPENDIX 3.1: WOMEN'S WASTELAND DEVELOPMENT GROUPS ORGANISED BY SARTHI

Year of formation	*Name of village*	*No. of women*	*Type of land members*	*Area (ha)*	*Remarks*
1987–88	Muvasa	29	Revenue wasteland	4	First group formed, excellent.
1988–89	Vandariya/ Kenpur	18	*Gauchar*	15.53	Still a bit weak due to membership from two far apart villages. Good protection of plantation.
1988–89	Gara	21	Revenue wasteland	11.35	Good potential, planted in 1989, discontinuity in guidance due to sudden departure of FC In-charge.
1989–90	Sadavra	16	*Gauchar*	4	Inspired by Muvasa. Still consolidating. Absence of strong woman staff member a constraint.
1989–90	Gamri	30	*Gauchar*	4	-do-
1989–90	Anjanwa	22	*Gauchar*	20	Weak group, Conflicts due to unauthorised encroachment.

(Appendix 3.1 Cont.)

Year of formation	*Name of village*	*No. of women*	*Type of land members*	*Area (ha)*	*Remarks*
1989-90	Vankadi	26	*Gauchar*	11	Excellent potential.
1989–90	Charad	52	*Gauchar*	38	Not fully formed yet due to interference by *Sarpanch*. Constraint of inexperienced FC In-charge.
1989–90	Padri	36	Revenue wasteland	10	In process of formation. Good leadership by woman staff member.

Total:

Number of groups formed : 9
Number of women members : 250
Number of ha : 117.88 (291.28 acres)

4 *Bringing Government and People Together: Forest Protection Committees of West Bengal—Role and Participation of Women*

PHILIP VIEGAS
GEETA MENON

INTRODUCTION[1]

Consequent upon two significant realisations—the scarcity of natural resources and the need for their judicious use—the management of India's natural resources has attracted much attention. As far as forests go, their history of management goes way back to the mid-nineteenth century when (in August 1855) a memorandum was drawn up by the fledgling Forest Department under the British India Government. It provided guidelines for restricting the rights of forest dwellers in order to conserve the forests. This was perhaps the first written document manifesting a changed attitude in the management of forests in India and indicating a shift towards the nationalisation of forests or what Singh (1987) terms as the 'privatisation of common property' through a subsequent plethora of forest legislation.

The country's forest lands are today officially and completely under Government control, managed by the states and governed

[1] We deeply appreciate the cooperation and support of several persons who made this study possible. In particular, we would like to thank U. Banerjee, A.K. Raha, A.K. Jha, S.N. Roy, S. Banerjee and several others from the Forest Department, West Bengal; S.B. Roy (IBRAD), B. Chaudhuri (the University of Calcutta), N. Banerjee (CWDS) and Nandini Sengupta and Sapan Das (RDA). We also owe our gratitude to all the members of the FPCs that we visited.

by the rules and stipulations of the Forest Acts and the National Forest Policy. The forests are, for functional reasons, divided into: (*a*) the Reserve Forests, which are given the highest degree of protection and have great commercial value, (*b*) the Protected Forests, which primarily play the role of environmental conservation, and (*c*) the village or unclassed forests, which are left to the use and management of the local population. However, given the commercialisation of forest resources due to the increasing raw material value of timber, softwood, etc., there has been, in most states, a gradual transfer of unclassed forest lands to the category of Reserve Forests. This was observed in Orissa (Fernandes et al., 1988) and in Madhya Pradesh (Fernandes et al., 1985).

Consequent upon this and other related processes, the management of available and accessible forest resources has been gradually slipping away from the control of the local tribal and other forest-dwelling communities so much so that most of these populations are today undergoing tremendous strains in their struggle for survival. This trend of gradual transfer of management of forests is visible even in the definition of the rights of the communities mentioned. The Indian Forest Act (1927) converted these traditional rights of the forest-dwellers into 'rights and privileges' and by 1952, these had been turned into 'concessions.' (National Forest Policy, 1952).

The increasing demand for forest raw materials that are associated with the process of industrialisation has led to large scale denudation of the nation's forest wealth. There is sufficient proof to indicate that the pace of denudation of forests in the country has been increasing steadily. It is estimated that on an average, the country loses 1.3 million hectares of forest cover every year (CSE, 1985). Though this process itself is alarming, what has been generally ignored in the debates on deforestation and its impact is the management aspect. This is our major concern here.

Prior to the commercialisation of forest resources, forests were managed in small patches by local rural communities. Most of them depended on these resources, ranging from food to medicine, for survival. The association of the Indian people with forests was so intimate that Indian culture was known as *aranya samskriti* and forests as *aranyani* (Bandhyopadhyaya and Shiva, 1987). This association was particularly intimate among tribal communities which depended substantially on forests for their economic, social and religious needs. Among the hunting-gathering economies, food came from fruits, nuts, edible leaves, flowers, roots, tubers, etc., and from the meat of animals and birds. In hunting-gathering and slash-and-burn economies, a rich variety of food crops was added to this. Besides food, forests were the sole source of energy for cooking and heating purposes, and provided material for building, fencing and other protection measures. Medicinal requirements were met from herbs and roots. Even in entertainment, forests played a major role. Given this relationship of intimate dependence, tribal culture in most areas was primarily oriented to preserving these life-giving and sustaining resources. Evidence of this can be derived from examples of protection and sustainable utilisation measures from all over India. Below are cited some of these measures:

1. The Chota Nagpur tribals scrupulously conserve a sacred grove (known as the *sarna)* as a symbol of communal identity.
2. *Pars* do not cut any tree in the sacred forests situated by the side of revers and streams.
3. Among the *angami nagas* it is customary for the users to plant and grow the number of trees that they cut (Roy Burman, 1989).
4. The Bhils name their *gotras* after the trees they worship. For example, the Jamnia Bhils worship the *jamun* tree, the Semalia Bhils worship the *semal* tree, the Palasia Bhils worship *palas,* the Anwalia Bhils worship *anwalia* and the Kanvi Bhils worship *aalam.* At marriages and other occasions the cutting of these trees is totally forbidden (Saxena, 1986: 38).
5. Before commencing slash-and-burn cultivation, most tribal communities offer *puja* to the local forest goddess to atone for the major offence of cutting and burning the forest patch for cultivation.

Besides these practices, several tribes have customary restrictions on the manner and extent of the use of forest produce.

The above, besides reflecting a certain attitude towards and concern for forests, are also examples of an informal management system. An exclusive feature of this system was that it was woven into the very religion and culture of the tribals. This lent to it a forcefulness even more powerful than the legislations of a formalised legal system since the latter function as impositions from above while the former constitute part and parcel of the total life system of a community. This is so precisely because it emerges from within the community itself, firmly rooted as it is in its existential experiences.

An important aspect of tribal economy in most tribal pockets of India is its women-centredness. In hunting-gathering communities, gathering was the main activity which ensured a sustained food supply. Even today, where it exists, it is primarily a woman's task. In slash-and-burn economies, the task of food production is mainly a woman's responsibility, regardless of whether the community is advanced or not. The main tasks of men were hunting, protection and warfare. Given this fact, it seems natural that women played an important role in the informal management system reflected in the tribal cultural ethos.

With the transfer of control over forest resources from the community to the Government, much of the essence of the traditional management system was lost. To this process is related denudation itself. Scarcity has led to the erosion of the traditional values which fostered preservation with the result that in most areas, the tribals themselves have begun cutting the very trees that they preserved in the past. Evidence of this is available from Orissa, Madhya Pradesh, Karnataka, Maharashtra and a few other areas (Fernandes et al., 1985; 1988; Gadgil and Malhotra, 1980). The net result is an increased pace of deforestation. The formal state mechanisms being primarily oriented to revenue generation rather than resource conservation, the latter aspect has been generally lacking in the overall management of these resources. Hence, scarcity of these forest resources experienced by the local tribal populations is closely linked to mismanagement by the formal management system.

The worst affected by this process are the local tribal communities. Deforestation has meant a sudden decline in availability of food. First, in terms of quantity itself there has been a decrease in the availability of vegetable foods (fruits, nuts, leaves, flowers, roots and tubers) and animal meat. Scarcity of forest land has led to intensive cultivation of smaller areas for slash-and-burn at a shortened cycle leading to soil erosion, which has further affected crop production. Besides food, the availability of medicines and housing material has also declined.

In view of the tremendous economic responsibilities of women, they have borne the main brunt of this degradation process. Investigations in Orissa have shown that women today have to walk greater distances to collect the increasingly scarce produce as well as strain themselves harder on the slash-and-burn land to produce whatever is possible on the degraded hill slopes. The depreciated nutritional intake and the reduced availability of local herbal medicines coupled with this increased intensity of labour have resulted in a steep decline in the health of women (Fernandes and Menon, 1988).

To this may be added the impact of the process of 'privatisation of communal property' on women. The main alternative available, and to some extent accessible, to the traditional economy is the settled agricultural system which, as it is generally practised, is based on male ownership and inheritance of land. The impact of this system has been so great that today, in many slash-and-burn areas of eastern India, even slash-and-burn land is individually owned, distributed and passed on in the male line. This has meant the complete removal of control over productive resources from female hands. Investigations in one block in Orissa have shown that this has led to women becoming completely dependent on men for their survival (Menon, 1989).

The impact on men has been equally significant though in a different manner. The main economic roles to be affected by forest denudation and the transfer of administrative control to a formal system were the male ones of hunting, warfare and protection. The loss of these roles has, as far as available evidence suggests, affected also the male's status. In some areas this loss of status was, along with economic scarcity, a major reason for the search for alternatives like migration (ibid.).

What, however, is significant for us to note at this juncture is the fact that along with this socio–economic degeneration the tribals and forest-dwellers, particularly the women who were visibly more affected than the men, were effectively kept away from participating in the management of the forests. The forests thus became the sole monopoly of the Government through the Forest Departments and Corporations (Kulkarni, 1989) while the forest-dwellers began to be viewed as 'enemies' of the forests from whom the forests had to be protected.

In stark contrast to this general attitude of the Government, the

Approach Paper to the Eighth Plan, however, appears to recognise the fact that in order 'to sustain social and economic development, especially in rural areas, steps are necessary to protect the remaining forests and to enhance the biomass resources....'

It further admits its errors with reference to forest management, and lays down that:

> sustainable management of forests would require an institutional framework which would facilitate people's active involvement. Traditionally, management of forest resources has sought to exclude the people, and the emphasis was on policing the forests to prevent biotic interferences. Predictably, this did not have the desired effects and has alienated the people. This has been particularly true in respect of the tribal populations who have been traditionally dependent on minor forest produce. Indeed, they have a national interest in protecting trees. Every effort needs to be made to promote grass-root level participation in this task as part of a larger approach to the local area planning and development. (Planning Commission, 1990: 44–45)

The Forest Protection Committees (FPCs) of West Bengal

Against this background, the phenomenon that is taking place in the forest regions of West Bengal assumes singular significance. The emergence of the concept of FPCs is striking deep roots in the Forest Department of West Bengal and in the hearts of forest-dwellers. The forest regeneration and protection movement is catching on particularly in the southern districts of West Bengal and FPCs are springing up all over. By 1990 there were around 1300 FPCs spread over the three southern districts of Purulia, Midnapore and Bankura, protecting over 1520 square kilometres of predominantly *sal* forest. West Bengal has an overall area of 88,460 square kilometres of which, according to data provided by the National Remote Sensing Satellite, 13 per cent or approximately 12,000 square kilometres, come under forest cover.[2] In

[2] Even this figure may be misleading as the interpretation of satellite data can be tricky in that it makes hardly any distinction between tree cover and mere shrub greenery.

the light of this, therefore, the contribution of FPCs to the forest cover seems highly significant.

Forest protection as a conscious move is not entirely a novel phenomenon. As most tribes resided in deep forests, protection of village forests formed a part of territorial protection and the tradition of warfare. However, protection of forest tracts was not an expression of the need to protect forests for their own sake. Rather, it was more an expression of the need to preserve the territorial integrity of the village. Hence, what is unique about FPCs is the conscious effort to revive a dying tradition with the intention of regenerating the natural wealth of the country. This in itself is significant in view of the economic, social and cultural deterioration of the tribal population and the environmental degradation that is affecting the climatic and soil conditions in the country. The second significant aspect of the FPCs is the managerial one. It refers to the collaborative attempt that consciously moves away from the general tendency of the government to gain complete monopoly over natural resources and the exclusion of the local population from participating in its management. The third important feature is the more recent one of the conscious effort to involve women in the formal structure and process of decision-making. This in itself is significant in view of the fact that women, though powerful in their economic roles as food producers and providers, have been traditionally denied access to participation in formal structures, especially those that take important decisions regarding the village and its members.

Though analytically these three dimensions are inseparable, the last one forms the major concern of this paper. What follows is a case study on the present and potential involvement of women in the FPCs, both at the formal as well as the informal levels, and the impact of the FPCs on their work and status. The FPCs attracted our attention first because in comparison to forest regeneration through the popular mode of afforestation, this was an attempt at regeneration through an entirely different method—protection. Second, the FPCs have been almost exclusively managed by men and though the Forest Department and voluntary organisations have shown eagerness to involve women in this management, there were several factors impeding this process. The analysis of these thus became important.

With this focus in view it was felt important, first, to visit as many FPCs as possible and to hold in-depth discussions with the members, particularly the women. Accordingly, ten FPCs operating under different conditions were visited. These ten FPCs are spread over two Revenue districts—Bankura and Midnapore—and fall under the two Forest Divisions of Bankura (South) and Midnapore (East). Originally, a tentative schedule was drawn up to visit the FPCs falling exclusively within the operational area of Nari Bikash Sangh (NBS) in Bankura district since a major thrust has been made by NBS to seek the maximum participation and involvement of women members of the FPCs. However, due to the prevailing political situation at the time of the study, and the Forest Department's restrictions regarding researchers visiting the area, our original schedule of field visits had to be modified to suit the situation. This made it imperative to solicit the support of the Forest Department which, in reality, meant a great deal of dependence on the local forest officials as an alternative. As may be obvious, from the methodological point of view, this approach had its own constraints. The major one stemmed from the fact that the forest officials accompanied us to the sites and were present during the discussions. However, our fears that their presence might influence the responses and reactions of the people as well as affect the direction of the discussions were allayed when we realised that the people had gained confidence and had achieved a certain amount of freedom to speak openly and spontaneously. In the selection of the villages, too, certain precautions were taken to minimise the influence of the local forest officials. Several criteria were laid before the Range Officers on the basis of which the villages were selected. Since the primary objective of our case study was to observe the active participation of women in the FPCs, it was imperative to the study to select villages that had the maximum involvement of women. It was also important by contrast to take up villages where there was little or no involvement of women; hence some FPCs having no women's involvement were also selected. It was felt necessary to add a further dimension to the sampling—that of the voluntary sector. Does the voluntary sector make a difference to the functioning of the FPCs in general and to the participation of women in particular? For this reason, five FPCs—Korapara, Balia, Juktidih, Pattarkati and Badashali—that have the direct support of local voluntary organisations were selected. Within this region Juktidih contributed a new dimension to the study. This village falls outside the administrative and political structure of West Bengal into that of Bihar but is geographically contiguous to the FPC region. Unlike West Bengal, the Bihar State Government has not yet issued any notification granting official recognition to the FPCs. This factor, we felt, would enable us to study by contrast the impact of the official State Government recognition on the functioning of the FPCs and the contribution of women participants.

Another major dimension that is of crucial significance for the state of the forests and their management is the factor of demography. Increasing population pressure of both humans and animals has often been shown to have a distinct negative impact on the state of forests. In this context and in the light of the fact that demography is closely interlinked with topography—both of which affect the feasibility of effective protection—we found that it would be essential to the study to include these factors as well. For this reason three villages—Tiring, Kattar and Khairbani—in the Khatra range were selected. The Khatra forest range is largely a low plains area as against the higher hilly areas of the Ranibandh range. The population density of the Khatra range is 752 per square kilometre while that of Ranibandh is 712 per square kilometre. The per capita forest area works out to approximately 0.24 acres in Ranibandh and 0.14 acres in Khatra.[3]

The Emergence of Forest Protection Committees

The concept of FPCs has its origin in the Arabari experience. Arabari is a small village some 20 kilometres from Midnapore town and falls within the East Midnapore Forest Division. In 1971–72, A.K. Banerjee, the then Divisional Forest Officer (DFO),

[3] This is a tentative estimate worked out on the basis of data provided by the respective Range Officers of Khatra and Ranibandh ranges. Khatra has a total geographical area of 93 square kilometres and a forest area of 39.02 square kilometres, i.e., 42 per cent, while its population is approximately 70,000. Ranibandh range, on the other hand, covers a geographical area of 52 square kilometres, and a forest area of 35.51 square kilometres, i.e., 68 per cent. Its population is approximately 37,000.

after repeated and concerted efforts at regenerating and protecting the forests, realised that because of the heavy dependence of the forest-dwellers on the forests, it was practically impossible to protect the forests without the support and cooperation of the people themselves. It was observed that if people's basic requirements of food, fuel and fodder were met through the restoration of their *nishtar* rights over forest produce, their cooperation would be forthcoming. Thus emerged a new concept in forest management. Whereas formerly the Forest Department took it for granted that the management of the forests was exclusively its own duty, this rethinking opened up new avenues for joint management wherein people's rights as well as duties towards the forests are given adequate recognition. Fundamentally, it reflects a radical change in attitude on the part of the Forest Department.

Banerjee realised that if people's cooperation was to be solicited, there was no other way but to take them into confidence. He, therefore, proposed his idea of joint management of forests to the people and thus initiated a dialogue with the people of Arabari village. The proposal made by their Forest Officer was acceptable to the people as they had as much to gain as the Forest Department by way of forest produce and small timber. It was acceptable to them also because it was in support of and in consonance with the attempts already made by the villagers to prevent encroachers from depleting their natural resource base. As a further incentive, they were also promised one-fourth of the income from the sale after a ten-year period when a yearly rotation felling of ten per cent of the area would commence.With this implicit pact between a forest official and the villagers of Arabari began an experiment that was to have a wide-ranging impact on the very concept of forest management. Based on this agreement, the first FPC in the region, that also included members from six neighbouring villages covering a total area of 1272 hectares of predominantly *sal* forest, was formed (Chandra and Poffenberger, 1989). The experiment went on for nearly a decade and a half without anybody ever taking notice of it till 1986, when a formal proposal was drafted by a group of interested Forest Department officials. Based essentially on the principles abstracted from the Arabari experience, this proposal sought to bring another 1,000 hectares under forest cover through the regeneration-cum-protection method. It proposed to undertake felling on a 10-year rotation basis and reward the 'protectors' with a 25 per cent share in the income.

It was estimated that a 100-hectare plot would yield an income of approximately Rs. 2.2 million of which the FPC would be entitled to Rs. 0.55 million. This was initially unacceptable to the Finance Department as it was seen by it as a loss of 25 per cent. Negotiations were initiated at several levels including the Ministry of Environment and Forests. At these negotiations the more positive points were highlighted and it was emphasised that the Forest Department would not be 'losing' 25 per cent of its income; on the contrary, it would be adding 75 per cent to its revenue which in any case it loses every year due to the absence of adequate forest cover. Gradually, the idea gained acceptance in official circles and an official order was passed in July 1988. In the meantime, however, the idea of the FPCs had become acceptable to the local people and, through informal communication, had spread even beyond the borders of Midnapore district to the neighbouring Bankura and Purulia districts.

Official Recognition and its Implications

Although the FPCs have been in existence for at least a decade and a half, the Government of West Bengal gave official recognition to the FPCs through its Resolution No. 4461-For. /D/15–16/88 on 12 July 1989. The specific recognition of the FPCs is expressed in the following paragraphs:

> Where as active participation and involvement of local people are vital for regeneration, maintenence and protection of aforesaid forest/plantation and successful implementation of the programme;
>
> NOW. THEREFORE, the Governor is pleased to decide that Forest Protection Committees shall be constituted for this purpose and beneficiaries acting as members of such committees shall be allowed, as a measure of incentive, 25 per cent of the usufructs subject to observance of the conditions provided in the Resolution.

The Resolution further goes on to delineate the composition, duties, functions, usufructuary benefits and restrictive measures pertaining to the FPCs.

Decentralisation and Joint Management

The Resolution essentially means the official recognition of the role of the people, i.e., the tribals and other forest-dwellers, in the joint management of the country's forests. In effect it abandons the understanding that forests are a monopoly of the state and that they are to be 'protected from the people' and accepts the idea that the people are equal partners and joint managers of forests.

The introduction of decentralisation in the management of forests has had a salutary impact on the people themselves. It is reminiscent of the old *aranya samskriti* of the country, when the people themselves preserved and even revered the forests. As we shall see later in this paper, the forest-dwellers have felt rather emboldened by this step of the Government. They have gained greater confidence and their chronic sense of helplessness and despair is being gradually replaced by a sense of hope and optimism. Socially, economically and politically they are beginning to feel empowered. A sense of belonging to and being rooted in their natural environment is once again taking shape within the community. Thus the official recognition has become a source of renewed hope for a new life that is emerging in these villages.

If the most significant contribution of the Government Order were to be singled out, it would most certainly be the empowerment of the local people. In other words, it is the creation of an atmosphere most conducive to people's active participation in the joint management of forests through actively empowering them in the true spirit of democracy. In the FPCs visited within West Bengal, the people felt that they were powerful now. As Rasi Mudi, a Santhal woman leader of the Korapara FPC and a member of the NBS, remarked: 'It's not just the Government or the Forest Department that is powerful. We too are equally powerful'. Many others also felt the same. In several of our discussions, the members claimed that because of the official Government recognition, they could fearlessly and authoritatively apprehend offenders, which prior to this recognition they were reluctant, and to some extent, unable to do as they had no legal authorisation. The official Resolution of the Government of West Bengal grants them this authorisation, thus strengthening and empowering the villagers.

By contrast, the impact of forest protection without this official authorisation is seen in the case of the Juktidih FPC. This village falls in the Ghatsila block of Singhbhum district of Bihar and, therefore, has no empowering mechanism like the Resolution of the Government of West Bengal supporting similar efforts of the FPCs within Bihar. As a result unlike the other FPCs in Midnapore, Bankura and Purulia districts of West Bengal, this FPC is struggling to survive. Although its formation has been spearheaded primarily by the local women with the encouragement and support of a voluntary organisation, the Rural Development Agency (RDA), protection of forests as such is proving to be a rather formidable task. Forest offenders are far too many, and with practically no Government support, the members of the FPC, particularly the women who are the more active members, are finding it increasingly difficult to control them. The Forest Department officials have, time and again, been informed about the illegal felling that goes on in these 'protected' areas, but apart from some assurances from the officials, no action of any consequence is taken. As many as three deputations consisting mainly of the local women members have visited the DFO apprising him of the regular forest offences. But once again nothing was achieved beyond a few assurances. On another occasion the women members intercepted 38 cart-loads of wood and timber that were being smuggled out of the forest. On being informed, the forest officials filed cases against these offenders but to date no punitive action has been taken against them. In the meantime illegal felling continues unabated bringing to naught the efforts of the FPC.

Politicisation of an Apolitical Structure

Along with decentralisation, however, there is an element of centralisation which seems to be introduced as a balancing mechanism in the structural composition of the Executive Committee (EC) of the FPC. Clause 1 (*v*) (*a*) specifies that the '*sabhapati* or any member of the Bon-o-Bhumi Sanskar Sthayee Samiti of the local *panchayat samiti*' shall be an ex-offficio member of the EC. Similarly, Clause (*b*) of the same section provides for the inclusion of the *gram pradhan* or any member of the local *panchayat* in the EC.

Here, perhaps, lies the crux of the problem of politicisation of an apolitical structure. The FPC is essentially a non-political structure. It is an element within a people's movement geared towards the protection of the State forests. Like the FPCs, the Forest Department too is a politically neutral organ inasmuch as it is a part of the bureaucracy and the administrative set-up. Tying the FPC to the local political structure, i.e., the *panchayat samiti*, through the imposition of the *sabhapati* and the *gram pradhan* (or any member of the *gram panchayat*) as members of the EC is tantamount to politicising an essentially neutral organisation. Although one may argue about the political neutrality of the elected body of the *panchayat*, in practice, however, one cannot ignore the significance of the *panchayat* elections for the State and indirectly even for the Central ruling political set-up. *Panchayat* elections are in a sense the barometer of the grass-roots base of a political party. Consequently, the inclusion of the *sabhapati* and the *gram pradhan* as *ex-officio* members of the EC is tantamount to politicising a non-political body since their very inclusion introduces an element of State control.

The reaction of the FPC members and the other elected members of the EC to the *ex-officio* membership status of the *sabhapati* and the *gram pradhan* is varied. In some villages, as in Tiring, Kattar, Khairbani and Patterkatti, people are not yet fully aware of this clause, or for that matter, of the other details of the Government Resolution itself. In general, however, when they were specifically asked about it they did not seem to resent it but expressed certain reservations, saying that the *sabhapati* would have to function like any other member on the EC with no special prerogatives by virtue of his office as the *panchayat samiti* head. They emphatically indicated that his views, ideas, etc., would be discussed on the basis of equality and within the democratic framework of the FPC forum. While stating at the same time, that thus far the relationship between the *sabhapati* and the other members has been one of camaraderie and friendship, the members did not rule out the possible eruption of a conflict in the future, particularly in view of the often diametrically opposed political affiliations of the *panchayat samitis* and the local villagers. While the latter are mostly staunch supporters of the Jharkhand Party, the former are chiefly controlled by the Left Parties. This was the predominant political atmosphere prevailing in practically every area that the study team visited.

The issue, however, stood out in bold relief in Balia village of Midnapore district. In this village, with the support and leadership of the RDA, an FPC was constituted in 1983. With the local Legislative Assembly and Parliament elections in 1985, the activities of the FPC received a setback. In a bid to woo the voters, the contesting politicians belonging to the Congress (I), the CPM, or the Jharkhand Party, encouraged tree felling in the forests protected by the FPC. The neighbouring member villages, therefore, under the cover of political patronage freely indulged in cutting trees for personal and economic gains. This had a highly demoralising impact on the people of Balia so that even the members of the FPC themselves, particularly the male members, could not desist from giving in to this temptation. Under extreme political pressure the RDA felt compelled to abandon the village organisation and move out prematurely. The net result of all this was that the membership of the FPC declined till eventually, for all practical purposes, it broke up and became dysfunctional. Today in the village there are just a few members, exclusively women, who are still keen on reviving the Committee. Balia, thus, is a clear case of political interference.

The Forest Department expressed its inability to tackle political 'interference' in its bureaucratic set-up. When questioned about it and about the potential politicisation through the inclusion of the clause regarding the *ex-officio* membership of the *sabhapati* on the EC of the FPC, a senior officer expressed total helplessness and stated, 'We are only officers!' Based on these undercurrents and given the power the political forces wield in some parts of West Bengal, it may not be too unrealistic to make projections of a difficult future regarding the trend of growth and manner of functioning of the FPCs in such areas. Further, giving these unhealthy trends legal status through Clauses 1(*v*) (*a*) and (*b*) in the Resolution is only adding more dimensions to the problem. Such interference undoubtedly has a direct bearing also on the role and participation of women in the FPCs because of the dynamics involved in the local politics which is almost exclusively controlled and dominated by men and, therefore, is generally out of bounds for women. This has been amply demostrated by the Balia instance cited above. Had it not been for political interference, the Balia FPC would perhaps have effectively protected a significant forest area through the participation primarily of the women of the village.

Representation of Women in the FPCs

Membership in the FPC is restricted to one representative from each household. It could either be a male or a female. Female membership ranges from a bare five per cent to as much as 63 per cent. [4]However, in most cases male representation invariably seems to be the norm. This owes itself mainly to two reasons: first, the tribal traditions that confer on men the responsibility of protection as an active measure. As mentioned earlier, forest protection formed a part of territorial protection, which was solely the task of men. Hence, it is assumed that protection in the form of policing should be a man's job even in the present context. Second, in today's context, this may also be tied to the internalisation of the dominant values which ideaise women as submissive and non-aggressive. Hence, the first reaction observed in villages where women's active participation was sought was that of surprise. It did not even seem to strike anyone that women too could officially represent a household as a member of the FPC. The committees are, therefore, predominantly represented by men. On the other hand, once suggested, no direct opposition to the inclusion of women representatives on the committees was observed in any FPC. On the contrary, in areas where this realisation has been creeping in, men have voluntarily surrendered their membership to make room for women representatives. Women's involvement has been accepted as a significant contribution to the efficient functioning of the FPCs. The Korapara FPC may be cited

[4] Accurate details on this are not yet available since the FPCs are still in the process of registering and updating the records as per the specifications of the Forest Department, West Bengal. The data that could be collected during our discussions regarding these figures should be considered more as approximate rather than accurate.

as a case in point. This FPC has the strongest contingent of women representatives. Of the 35 members, 22 are women and as we were told in the discussion, there are moves afoot to increase the female representation even further. The women in this FPC feel that they have a 'very major role' to play in the protection of the forests. Initially, at the formation of this FPC in 1986, it had exclusively male members. But gradually with the intervention of the NBS, motivation was generated for greater female participation in the village activities in general and in the FPCs in particular. Similarly in Juktidih too, it is the RDA that is largely responsible for the increased female participation in the FPC. The RDA has a team of three members working in the area, one of whom is a lady health worker. Prolonged work with the women has helped raise their consciousness which is why there is greater female representation in these FPCs. The 30 per cent female representation in Maknu village may primarily be attributed to the circular passed by the DFO, Bankura (South) Division, which stipulates that there should be a minimum of 30 per cent women representatives on the FPCs. This stipulation is welcomed by the Forest Officers of the Division, who are now seriously attempting to implement it.[5] Concerted efforts, therefore, are now being made to include as many women as possible in the FPCs.

The stipulation, it must be admitted, will go a long way in improving female participation in the FPCs and in the joint management of the forests. However, the rationale for promoting greater participation of women in the FPCs reflects more a pragmatic attitude than a desire for raising the social status of women. A very practical reason underlies this step towards extending direct membership to women. It was found that in most areas the maximum number of forest offenders were women coming mainly for the collection of fuelwood and non-timber forest products (NTFP). Even the cutting of *sal* trees for poles for sale was done by women while their husbands did the actual sale in the market. Further, it was observed in several areas that, consequent upon the formation of the FPCs, there was an increase in the number of female offenders corresponding to a decrease in male offenders. It was pointed out in this connection that men urge the women to encroach on the forests because of the fact that apprehending

[5] This Resolution was passed after the NBS wrote to the Forest Department in this connection.

female offenders was a job that the male representatives of the FPCs, as well as the Forest Department functionaries, found difficult to accomplish. Hence, it was felt that since women informally protect the forest from these encroachers, having women members on the committee would motivate them and facilitate effective protection. Therefore, it seems that, more than humanitarian considerations of equality and social status, it was the gender-specific role based on sheer necessity emerging from the inability of the men to handle the situation that led to the promotion of women's representation in the FPCs.

In areas where voluntary organisations like the NBS and the RDA are functioning, the above fact was combined with that of the awareness that women should be equal partners in what essentially concerns them in view of their economic responsibilities. This was particularly so in the Korapara FPC where the NBS is operating and where this awareness was clearly expressed.

Where women's representation is low, as in Khattam, Tiring, Khairbani, etc., the contribution made by the women members at the meetings seems small and insignificant, even though they have equal status within the democratic structure of the FPC. The domination of males in terms of numbers perhaps acts as an inhibiting factor whereby the women feel reticent and diffident to express themselves freely and spontaneously. In many areas the women did express interest and a keen desire to be active members, but they felt that they were too overloaded with household chores to spare time for the committee meetings. They also seem to be victims of the 'culture of silence'. In some other areas the women felt more at ease sharing and raising the issues with their husbands at home rather than at the open meetings, and felt assured that their husbands would represent their views at the meetings.

The Unrecognised Protectors

In complete contrast to their representation in the committees, the active contribution of the women to the aims and objectives of the FPCs is by any standards much more than that of the male members. This seems to be the irony of the situation in that recognition is given through official membership to the men whose contribution is much less than that of the women. In view of their economic responsibilities, it is the women who visit the forests regularly. As one of them observed, there is no time in the day when a few women are not present in the forest. Hence, in most areas no need is even felt to appoint special patrols to guard the forests from offenders. The women invariably take on this role. Whenever they spot an offender it is they who apprehend him/her directly, although if the offender proves too much for them they report him/her to the other committee members.

The case for greater women's participation is strengthened by the experience of the Korapara FPC. Here the FPC is spearheaded by women. They feel sufficiently empowered and motivated to face any eventuality. Not only have they gained greater power but also greater consciousness through the FPC as is clearly reflected in a pertinent remark made by one of them: 'Earlier we thought free mixing of men and women was an offence—now we don't think so—we can and must mix freely if we are to solve common problems.'

The Balia FPC in Midnapore Forest Division too is led by women. Although it seems defunct, as explained earlier, a few women members are still trying to keep alive the consciousness generated by the task of protection. Their motivation is still high and they are still hoping that if the men give them a little support, they should be able to revive the committee.

The experience of Juktidih (Bihar) is even more compelling. Here the women hold the key to the leadership of the FPC. In spite of the lack of support, both from the male members and from the Forest Department, the women seem to be fighting a lone battle to protect the forests. Because of their commitment to the protection of forests, they even bear harassment from the local males in public places like buses and markets.

Benefits

In terms of forest regeneration itself, the West Bengal Forest Department's concerted efforts to promote FPCs have shown good results. The regeneration of *sal* forests through the FPCs appears very impressive in the southern districts. Statistics are still being compiled on the FPCs at the State Forest Headquarters in Calcutta. Available statistics indicate that at the end of 1989 there were about 1,300 FPCs covering about 1,520 square kilometres of *sal* forests. West Bengal has a total forest area of 4,500 square kilometres of which 2,500 square kilometres are under *sal* and capable of rejuvenation. Through further expansion of FPCs the Forest Department expects to regenerate and protect this entire *sal* area.

This regeneration of the *sal* forest has come as a great boon to the forest-dwellers, particularly to the women. Unlike other exotic species such as the controversial eucalyptus, *sal* allows the natural regeneration of other indigenous species and a variety of undergrowth. Apart from contributing a natural characteristic to the forests and enriching it with its biological diversity, this companion growth further improves the availability of NTFP on which the dependence of the forest-dwellers is extremely high.

With increase in forest cover there has been a general improvement in the availability of fuelwood, fodder and all the locally available NTFP including herbal medicines. Fuelwood has always been one of the prime concerns of women. The steady decline in tree cover has resulted in their having to walk long distances and, in some areas in India, even as much as 18 kilometres a day to obtain the day's quota of fuel energy.

Through the FPCs the women have certainly found a solution to the fuelwood problem and the associated anxiety. One example may be sufficient to highlight this benefit. The villagers of Patterkatti informed us that owing to the non-availability of firewood, the villagers were depending on the market, agricultural waste and the forests (in that order) for their fuel requirements. With no forests left in its vicinity, dependence on the forests meant even digging out roots which had thus far been preserved because of their immediate inaccessibility. With the regeneration of 78 acres of *sal* forests through protection, fuel requirements for four to five months of the year are now met through the biannual thinning of the regenerated *sal*. Agricultural wastes fill in for the remaining part of the year. There is almost no dependence on the market where one headload of fuelwood costs as much as Rs. 10. With one stroke, therefore, the excessive energy expenditure of the women, their anxieties, as well as the cash outflow of the household, have been minimised. The gradual return of NTFP in relative abundance and the restoration of their original rights of free access to them is another major source of satisfaction to the people. While just a few years ago there was a steep decline in the availability of NTFP, today the trend is being reversed. All varieties of NTFP, such as leaves, flowers, fruits, seeds, oils, resins, roots and tubers and medicinal herbs are now becoming more and more available. It is perhaps as yet too early to make an accurate assessment of their availability and of the impact it has on the local economy and household usage. However, in the village of Badashali, the village youth conducted a survey and estimated that their local forests, if protected, could produce NTFP worth about Rs. 100,000 annually. This motivated the villagers to go in for forest protection and controlled use of NTFP. Today, after five years of protection, the villagers claim that they have sufficient NTFP like mushrooms, *sal* seeds, *mahua* flowers and seeds, leaves for leaf plates and cups, different varieties of resins, medicinal roots and herbs, etc., for their personal use. This has helped the women tremendously. Although most of these are not sold for cash income, the women claim that their usage at home directly cuts down their expenses and work considerably. Other villagers in Maknu, Korapara, Juktidih, Tiring, etc., too made similar claims. Further, there was general agreement about the improvement of health in the village due to increased availability of herbal medicines. In Juktidih, the village lady health worker of the RDA was training the village women in the use of the locally available herbal medicines.

The potential for medicinal herbs as a major source of income seems very high and varied. *Kalmegh, parura, chaknni, patal gudu, ram datan, anantmool, sat mool, janggi, sona, lodh,* etc., are available in vast quantities. In Ranibandh range we had the occasion to visit the godown of a middleman who does business exclusively in the locally available medicinal herbs. These are bought by him at the local market rate and sold to other larger middlemen in Calcutta who have links with major pharmaceutical companies. According to this local middleman, the marketing of these herbs is financially so lucrative that a sale of just two or three truckloads of these each year fetches him enough income for the entire year. Subendu Banerjee, the Range Officer at Khatra, estimated that each FPC could earn a minimum of Rs. 50,000 annually on the sale of NTFP alone. Table 4.1, compiled from data provided by Sapan

Das (RDA), is indicative of the extent of the potential wealth in the forest regions. To obtain an exact estimate of the forest wealth would require a separate economic survey and analysis of each specific area since the NTFP availability of one area differs from another.[6] The regeneration of forests has thus generated a tremendous amount of wealth in the form of local natural resources which, if tapped scientifically, has a vast potential for raising both the economic and the social status of the local people.

Several factors, however, restrict the adequate utilisation of this vast resource. Personal consumption of these resources needs to be promoted in view of the impact it can have on the health of the people. However, improper marketing facilities limit the income-generating capacity of even saleable and surplus produce. These are either sold to the local buyers and middlemen or to the local Large and Multi-Purpose Society (LAMPS) at a very low and non-competitive price. To the former set of buyers the sale may be characterised as distress sale in the absence of better and easily accessible alternatives. For instance, *tassar* cocoons are sold to the local buyer at 10 to 15 paise for each piece even though the price offered by the Sericulture Department ranges from 25 to 30 paise per piece. This is so because the Sericulture Department is located too far away from the villages. (See Table 4.1)

To the latter set of buyers, the sale is due to regulations that have led to the monopolisation of these products by the Cooperative Societies. LAMPS, for instance, holds total monopoly over *kendu* leaves which are used for *beedi*-making. The herbal medicines too could fetch a much better price than is being offered by the local middlemen.

Apart from the lack of a proper marketing system, there is also a lack of adequate technology. The introduction of small scale and low cost technology can prove to be of tremendous benefit. For instance, the scope of leaf-plate making could increase by leaps and bounds if machines for the purpose could be made available to the people. Besides increase in income, it would also generate additional employment, especially for women who are the main workers in this 'industry'. Machine-made plates cost 25 paise a piece, while those made manually cost half as much. Initial investigations have shown that there is greater demand at present for machine-made leaf-plates than for hand-made ones.

[6] See Malhotra and Poffenberger, 1989:41–43.

Table 4.1

Details of Locally Available NTFP

Name of NTFP	*Purpose for which used*	*Quantity (one yr.)*	*Collection time*	*Local rate (Rs.)*
Sal : seeds	Oil	Found in abundance	March–April	3.00/kg
: leaves	Plates		Throughout the year	10–15/1000
Mahua : seeds	Oil	—	March–April	2.50/kg
: flowers	Liquor	—	Feb–March	2.50/kg
Kendu : leaves	Beedis	Plenty	Except monsoon	50.00/bag
: fruits	Eating	Plenty	March–April	1.00/5fruits
Arjun : bark	Medicinal	Few trees	Throughout	1.00/kg
Amlaki	Medicinal pickles	Very few	Dec–March	2.00/kg
Harataki	Medicinal	Plenty	Nov–Dec	1.00/kg
Mushroom : small	Edible	Plenty	June–July	4.00/kg
: large	Edible	Plenty	June–July	8.00/kg
Honey	Edible	Plenty	March–April	20.00/kg
Dhuna	Religious	Plenty	March–April	20.00/kg
Lodh : bark	Medicinal	4–5 kg/tree	Throughout	0.50/kg
Sona : bark	Medicinal	4–5 kg/tree	Throughout	0.50/kg

Table 4.1 (continued)

Name of NTFP	*Purpose for which used*	*Quantity (one yr.)*	*Collection time*	*Local rate (Rs.)*
Bacher : bark	Liquor	1 kg/tree	Throughout	15.00/1000 (pellets: size of marble)
Janggi	Medicinal	Plenty	Nov–Dec	2.00/kg
Kal megh	Medicinal	3–500 gms/tree	Sept–Dec	Freely available
Sat mool	Medicinal	1–200 gms	Throughout	2.00/kg
Anant mool	Medicinal	1–200 gms	Throughout	3.00/kg
Ram datan: root	Medicinal	Plenty	Throughout	0.50/piece (5–6")
Patal gudu: root	Medicinal	1–200 gms	Throughout	2.00/kg
Laha (lac)	Commercial	Plenty	Throughout	18.00/kg
Tassar	Commercial	Few	—	20–25 paise (piece)
Rahara: fruit	Edible	Few	April–May	No sale
Parura: bark	Medicinal	Plenty	Throughout	No sale
Chaknni: bark	Medicinal	Plenty	Throughout	No sale

By and large, therefore, these two factors—lack of adequate marketing facilities and of appropriate technology—may be singled out as the restricting factors limiting the scope of income and employment generation. The Forest Department has provided other support systems by way of supporting agricultural activities through irrigation facilities, etc., but as far as NTFP are concerned, apart from granting the people their usufructuary rights over them, no specific supportive measures by way of processing and marketing have been taken up. Nor has the Forest Corporation been of any help since its activities are restricted only to timber and timber products.

One other major benefit, as already mentioned, is that the people have been promised by the Forest Department a 25 per cent share in the net sale of the major produce of the forest—*sal* poles. It is estimated that at the current market rate, income from the sale of poles per hectare would be around Rs. 20,000, giving the FPCs a share of Rs. 5,000 per hectare. This could fetch the FPC a substantial amount of income every year once the 10-year rotation felling commences depending, of course, on the size of the protected area.

'Thinning' or pruning of the *sal* trees provides a large quantity of the much needed biomass for fuel from time to time. This takes place at least twice a year and each individual household gets a substantial amount of leaves and twigs as firewood. In Tiring, for instance, the thinning of the protected area in 1989 produced 344 cart-loads of leaves and twigs. One-fourth of this, i.e., 86 carts, were acquired free of cost by the villagers, while the remaining was bought by them at a cost of Rs. 20 per cart. In some areas the Forest Department has undertaken supporting activities in addition to the active promotion of FPCs. In Maknu, Pukuria and Khattam, for instance, it has financed small earthen dams. These dams irrigate the fields of the local farmers in addition to improving the water-table of the area. The Maknu earthen dam irrigates about 40 acres of the local lands throughout the year, thereby enabling the raising of two crops a year where only one crop was grown earlier. This naturally supplements their income from the forests. An additional source of income from these earthen dams is the promotion of pisciculture. We were informed by the villagers in Maknu that, with a minimum investment of Rs. 1,000 for fingerlings, they could have one harvest every six months worth around Rs. 25,000. The local villagers, who feel well supported by the Forest Department through these supplementary activities, coupled with the flow of other benefits they get from the forest, foresee a bright economic future for themselves.

These attempts of the Forest Department seem to be highly creditable since they are aimed directly towards the economic upliftment of the people. However, better results could be achieved if community building could be infused into these economic measures. For instance, the pisciculture project in Maknu will benefit only the 14 households who have invested some of their land and money in the project. The other members of the village have no share in this. Benefits could be extended to the other villagers if, for instance, a village fund for offering soft loans could be created by the FPC. Besides being of economic benefit it could also bring about social benefits. Giving it a greater community orientation would make a vast difference in the integrated growth and development of the village and in encouraging to some extent the community feeling which is known to have existed among tribals in the past.

Employment Generation

A major contribution of the FPCs has been the generation of employment for the local people. According to A.K. Raha, Additional Chief Conservator of Forests in the Forest Department, West Bengal, employment generation is the 'main thrust' of this programme. The following is a list of activities having a potential for the generation of employment that the Forest Department has undertaken in the villages:

1. Thinning
2. Multiple shoot cutting
3. Farm forestry
4. Silvi-pastoral farming
5. *Tassar* host plantation
6. Cashew plantation
7. Digging wells for drinking water
8. Constructing earthen dams

9. Clear felling
10. Village road construction

These employment-generating activities are supported under the 'Rehabilitation of Degraded Forests' Scheme. In essence the extent of employment generated, measured in terms of person-days, is largely dependent on the availability of funds for the particular project. For instance, the earthen dam in Maknu was constructed at a cost of Rs. 60,000. Setting aside an estimated cost of about Rs. 20,000 for material, etc., it leaves approximately two-thirds of the amount for labour. At the prevalent wage rate of Rs. 23.11 per person per day for manual labour, it generates a total of 173 person-days of employment. Thus, each of the projects has the capacity to generate employment, but only intermittently, under the inevitable constraints of availability of finance as Table 4.2 (compiled from data provided by B.N. Roy, Range Officer, Ranibandh) shows. It is clear from Table 4.2 that as far as daily wage labour, piece-rate and salaried employment in the forestry sector are concerned, the scope within the Forest Department is

Table 4.2

Estimated Employment Generation Potential of Forest-related Activities in Ranibandh Range 1988–89

Time frame	*Activities*	*Amount allotted*	*Employment generated (person-days*)*
First year:	Advance work for planting	Rs. 231.00	10
	Other advance work (pits, soil work, etc.)	Rs. 1,594.00	69
	Planting, raising seedlings guarding, etc.)	Rs. 2,250.00	97
Second year:	Maintenance, mulching, guarding, etc.	Rs. 1,600.00	69
	Multiple shoot cutting	—	11

* Figures are rounded to the nearest decimal.

very limited. By contrast, as we have already mentioned, as far as self-employment and 'unpaid' labour are concerned, the scope seems unlimited.

The processing of NTFP offers wide avenues for generation of steady and sustained self-employment patricularly as far as the women are concerned. These possiblities, however, need to be further explored since much is dependent on the availability of local resources and on the scope the particular NTFP offers for processing. In some cases, like *kendu* leaves, it may be limited to collection and sale, while in others like *tassar* cocoons, it could be very extensive right from the cultivation of *arjun* trees up to spinning of the yarn and weaving of the cloth as a full-fledged small scale handloom industry. This has particular application to the rural women. The actualisation of this potential, however, implies that local entrepreneurship will be strongly supported by a credit and marketing infrastructure that is presently found wanting in the area. The Forest Corporation could have been one of these agencies that could and should have provided this much needed support. But this would require a radical structural and administrative reorientation of the Forest Corporation. Nor are the local tribal cooperatives of any help, given their present manner of functioning. One alternative may be an 'external agency' that is able to inject the required appropriate technological know-how and the entrepreneurial spirit into the local economy. The feasibility of this needs to be looked into.

Social Benefits

Preliminary investigations have shown that, prior to the formation of the FPCs, *namaal* was the annual seasonal routine particularly for the males and generally to the neighbouring districts of Burdwan,Hooghly and Howrah. As a relatively widespread phenomenon, the incidence of migration appears to have been extremely high prior to the formation of the FPCs, as indicated by the residents of Maknu village where they informed us that seasonal migration used to be as high as 80 to 90 per cent. Forest regeneration has reduced this to the current level of 40 to 50 per cent. Although still quite high, the decline is quite substantial considering that forest regeneration and protection are just taking off.

Although no in-depth data collection was done in the villages that were visited, it is generally observed that both seasonal and long-term migration can cause wide-ranging changes in the local community owing to the introduction of new and alien social and cultural values. Preliminary investigations in the areas visited have shown that the main negative impacts of this process are those which are inflicted on women. Among other things, male migration leads to a tremendous increase of the work-load of women. Second, in view of the fact that traditional practices like polygamy still persist, women are subjected to the psychological anxieties of desertion and/or having to adjust to a co-wife. Though this issue can be a subject of another research study, initial investigations have shown the existence of both these problems. Hence, it is but logical that the reduction in *namaal*, brought about by the introduction of FPCs, is bound to lead to improvement in the work and lives of women.

Stronger village organisation, better unity in the village, greater political and social consciousness, ability to fight for their rights and to clearly articulate their demands are some of the other social benefits that may be attributed to the formation of the FPCs.

The women in particular have found these committees very helpful as they have been instrumental in generating mutual understanding and sensitivity. In Tiring, for instance, a young lady of 23 years of age, who has now been deserted by her husband, has received support and understanding fıom the other female members of the FPC besides financial help. As she has no land ownership rights and, therefore, owns no land, the organisation of the FPC and the resultant increase in sensitivity towards the struggles of each other has been of tremendous moral and psychological support to her. This demonstrates that the women now enjoy a particular intimacy and closeness with each other from which they derive a certain psychological strength and security. Solidarity is expressed by coming to the aid of the needy women such as those who are pregnant or have problems with their husbands. In Juktidih, for instance, some women reported that formerly while pregnant women had to fend for themselves, today they get support from others. If, for instance, a pregnant woman has to be brought to the clinic—some seven kilometres from the village—she is taken there by the village women. Such instances are only indicative of the deeper emotional ties of unity and solidarity that have developed in the course of the formation of the local FPCs.

More importantly, a strong sense of self-worth, dignity and belonging is beginning to strike roots among the tribal and forest-dwelling communities accruing from the fact that they are now the 'joint managers' of the forests. This was expressed by several of the members very emphatically. The sense of 'uprootedness' and displacement and/or dispossession is steadily giving way to this sense of rootedness and belonging.

Conclusion

In the context of the fact that West Bengal has the second highest population density (730 per square kilometre) in the country (Muthiah et al., 1987: 22) and given the related problem of intense poverty conditions prevailing in the State, an intensive land reform package is a necessary precondition to poverty alleviation and the promotion of social justice. However, land reform in its narrower definition of redistribution of land in favour of the landless and small cultivators and tenancy reform will not be sufficient to serve these ends. Since no more land is available for redistribution, the tribals stand little or no chance of benefiting from the land reforms package. In this context, the joint management of forests seems to provide a suitable alternative. In this sense, the land reforms package may be broadened to include joint management of forests where forest lands are renewable resources. If managed effectively through protection and product optimisation, they can substantively increase the income and living standards of the dependent population.

The Seventh Five-Year Plan clearly draws the linkage between the land reforms package and anti-poverty programmes when it states that:

> the core of the anti-poverty programmes lies in the endowment of income generating assets on those who have little or none of these. Hence redistributive land reforms and security of tenure to the informal tenants have to be directly integrated

with the anti-poverty package of programmes (Quoted in Datta, 1988: 4.).

In no programme is this integration made more effectively than in the FPCs which aim not merely at increasing the forest cover and the upgradation of degraded lands, but also at involving the forest-dwellers in the process of regeneration and at the same time contributing to their economic upliftment. In tangible terms, they have been officially granted full usufructuary rights in the protected forest areas as well as a 25 per cent share in the net sale of timber at the time of felling. These steps go towards substantially increasing the per capita income. Though ownership is only symbolically vested in the forest-dwellers, the feeling that the forests belong to them and that they have rights in them is in itself a significant achievement. The added incentives in the form of economic programmes like the construction of earthen dams, pisciculture, etc., would be indicative of a commitment to such an integration.

The State of West Bengal has a history of attempts at promoting joint management of land. This, for instance, was the driving force behind the relatively successful schemes like 'Operation Barga' (Datta, 1988: 5–10). Measures of political consequence like the FPCs, which link land reforms to rural development with an equal emphasis on the grass-roots organisations, further reaffirm the sincerity with which attempts at promoting equity and social justice are made. In this context, a significant question which may arise is whether such attempts are linked to the political ideology of the left movement or whether it is also compatible with other political ideologies as well. In other words, is it merely an administrative question or is it also a political one? While this question cannot be answered in simplistic terms since no comparable collaborative attempts have been made elsewhere, the fact that a major alternative to the common top down approach has been initiated in West Bengal is in itself significant.[7]

The Forest Department of West Bengal was motivated by the loss of the native *sal* forest, the accompanying decline in forest

[7]It may be mentioned here that equally successful forest protection measures have also been taken up in States like Orissa mostly by the voluntary sector. Only recently are some tentative attempts being made by the Government of Orissa to promote joint management of forests with the village communities.

revenue and the need, therefore, to regenerate the *sal* forests. In the Arabari experience the Forest Department found the most appropriate paradigm for the achievement of these objectives. It was perceived as a suitable model of joint management of the forests.

Prospects of a brighter economic future have been the main motivating force for the people. Frustrated by the non-availability of fuel energy and fodder as well as by the steep decline in the flow of seasonal NTFP, they saw no better option other than the regeneration of their forests. A substantial share in the net proceeds from the sale of timber was an added incentive.

While this has been the general focus for the people as a whole, one could distinguish gender-specific differences in the emphasis given to the overall thrust of 'economic' benefits. The male members of the FPC, who were more attracted by the estimates of economic benefits highlighted by the Forest Department, exhibited a greater concern for items that have a higher economic value. Compared to this, the women were more emphatic about the availability of firewood and NTFP that are of household use—those that meet the food, fuel and fodder requirements. This indicates that the men's primary concern was the earning of cash income while the women's was the provision for and care of the family. The women's measure of motivation was distinctly higher than that of the men, indicating a relatively more intimate and intense relationship with the forests. In Patterkatti, for instance, the women told us that they would try and persuade the Forest Department not to cut the forest at the end of the 10-year rotation period. They were ready to forego their 25 per cent share in the net produce in order to protect the forest. One woman even said: 'We will never allow the forests to be cut.' Underlying this was the deep feeling of responsibility that the women felt both for their families and for their forests as indicated by the Balia village experience.

For decades, the women had struggled to provide fuelwood, supplementary food and other essentials for the family, regardless of all the physical and mental strain undergone in the process. With the establishment of the FPCs, these conditions are beginning to change. For instance, even as recently as three or four years earlier, the women had to cover great distances foraging for fuel and supplementary food items like fruits, nuts, leaves, etc.

Dependence on the market for these essentials was also on the increase, thereby causing great strain on their meagre finances. Today, both the physical and the economic strain have been greatly minimised since the required items, at least for part of the year, are becoming more easily available and accessible.

The second set of benefits may perhaps be seen as indirect benefits and are usually important for a population victimised by the dual factors of poverty and gender discrimination. These can be expressed in the re-establishment of the women–forest ties on the one hand and the women–women ties on the other. For tribal women, the establishment of the former brings back some of their lost status since in the past the socio-economic status of women was closely tied to their capacity to provide food, fuel and other essentials. A hard-working woman, who can best satisfy the needs of the family, was also the most respected and sought-after one.

The women–women ties that the FPCs have helped to re-establish is yet another major benefit. We have observed earlier that, with the growing competition and physical burdens, the time available to women to socialise or lend mutual support had been greatly reduced. They, therefore, were bearing physical and psychological burdens alone. Today, owing to the closer contacts brought about through the FPC experience, a woman can, to some extent at least, depend on her friends and the villagers in general in times of need. We have alluded to several examples of this.

Another major concern in this study has been the women's actual and potential roles in the FPCs and the factors impeding these. Our first conclusion in this regard is that women already play a major role in the activities of the FPCs. The traditional tribal male tasks of protection of territories, hunting and warfare, having been rendered redundant through the process of change, there has been a diversion of male labour from forest-related activities to settled agriculture, commerce and other ventures. The women, on the other hand, have had to retain their traditional roles due to the absence of viable alternatives to the provision of food, fuel and other necessities to which are added their modern ones related to settled agriculture, migration, labour, etc. Their on-going forest-related work, as we have shown earlier, requires their constant presence in the forests. Consequently, their participation in forest protection is easier and follows as a logical course of action. Precisely because of this, women's actual role in forest protection seems to be taken for granted. It is only through the government stipulations regarding female membership that formal recognition has been given to these indispensable tasks. This not merely adds an altogether new dimension to the issue of women's participation in forest management but also marks a new phase in the lives of tribal women who now have access to formal decision-making structures which were traditionally beyond their reach. This is a major achievement of the FPC where women are concerned. Hence, the decision to promote the representation of women in the FPCs is a laudable one in view of its consequences, even though the factors that have motivated the decision were more utilitarian than ideological ones relating to establishing gender equality.

Being a logical extension of their other economic and subsistence activities, women's informal involvement in forest protection is somewhat similar in all the areas studied. Their formal representation, however, varies with the operation of factors like cultural values, political awareness and motivation. In areas where the 'culture of silence' persists, women have been unable or perhaps dissuaded from shedding their traditional inhibitions in participating in formal decision-making processes. Where this disability has been overcome either through the support of voluntary organisations like the RDA and the NBS or through a direct awareness of the existing political situation and the need to participate in it, the involvement of women in the formal structure of the FPC has been very high, so much so that it has even exceeded that of men. A combination of the two can be found in several cases: for instance, in the Korapara FPC, where the already existing political awareness of some women has been given impetus by the activities of the NBS leading to a high female representation of 63 per cent in the FPC membership.

By contrast, in some areas like the Khatra range, the absence of all these factors has minimised the participation of women in the formal functioning of the FPCs. This is especially so, as mentioned earlier, because unlike the neighbouring Ranibandh range, the Khatra range has a higher population density thus putting greater pressure on the forests. This, combined with the fact that forests existing on such plain lands are more accessible to outside commercial interests, presents greater hurdles in forest protection and regeneration. This very fact of accessibility also implies greater exposure of existing tribal populations in the area to outside cultures, especially those that place comparatively more restrictions on women's freedom of movement and speech. The odds that these present against women's involvement in forest protection are in themselves great. Accordingly, we observed that in the predominantly Munda village of Tiring, women's informal participation in the FPC is high, but formal involvement very low.

In some areas, political factors have presented major hurdles in the participation of women in forest protection. In Balia, in fact, this force has threatened the very existence of the FPC, and compelled the local voluntary organisation to withdraw. This measure has further demotivated the women. Nevertheless, Balia is itself a good example of the potential of women in keeping the FPC movement alive, if the necessary external support is available. Women's concern regarding forest protection has been kept alive in this village both by the economic loss they experience and the fact that the FPC has created a certain awareness regarding the need for unity and the sharing of concerns. In view of this, the potential for rejuvenating the FPC is very high.

Juktidih and Patterkatti add a different dimension to women's potential in forest regeneration: the traditional dominance of women in the economic field, the men's relative 'backwardness' in this regard and the significant fact that women have surmounted their traditional inhibitions in participating in the formal decision-making structures and in public speaking have combined to make women a force to reckon with.

The role of the RDA in this achievement cannot be underestimated. It has been the main driving force behind it. The combination of these factors, if extended to other areas, can open up wide avenues for both forest regeneration and women's development. This is precisely what is lacking in the Tiring FPC in the Khatra range where the role of voluntary organisations needs to be maximised to achieve this goal more effectively in view of the seemingly insurmountable odds. Informal discussions in Tiring with women showed that despite their inhibitions, women's interests in formal representation matches their commitment to forest protection. The local leadership, while acknowledging women's present and potential roles, is unable to take the initiative in this. In most areas in general, the traditional equation of males with protection still persists. Here, therefore, lies a vast potential which can be tapped if the Forest Department which is attempting to promote women's representation is supported by the active involvement of NGOs.

The full achievement of the objectives of women's complete involvement and representation cannot, however, be achieved, nor the formidable forces against the very existence of the FPCs adequately understood without placing the FPCs within the context of the wider policy matters. The contents of the Forest Policy and Forest Act and the objectives of the FPCs seem to be in contradiction with each other. One officer aptly described this contradiction by colourfully portraying the forest official as a goddess whose one hand blesses the people with prosperity while the other restricts their rights. The contradiction in this picturesque portrayal seems quite apparent: first, while the very functioning of FPCs assumes a cordial relationship between the forest officials and the forest-dwellers, the exploitation and commercial viewpoint and the persistence of the 'policing' orientation inherent in the Indian Forest Act and the Forest Policy make it rather difficult for the forest officials, especially those who have to deal directly with the people, to extend this relationship beyond a certain point since they have to ultimately work within the general thrust of these policy measures. The general thrust of the forest legislations of the country still remains the promotion of industries at the cost of people's rights in the forest and the viewpoint that still holds is that the people are responsible for the present condition of the forests. Within the purview of this thrust a forest official committed to the promotion of joint management and people's power encounters an ideological contradiction that is difficult to reckon with.

Second, and related to the above, is the lack of delegation of power to the lower ranking officials like the Range Officers. For instance, as of now, all the decisions within the Division are centralised in the DFO. The Range Officers have to seek his permission even to grant a few logs of wood to the people for funeral rites which, given the distances and lack of communication, becomes difficult and impractical particularly in view of the nature of the emergency. This obviously does not help promote any rapport between the Forest Department and the people.

Hence, radical and extensive amendments in the Forest Policy and Act are important requirements for neutralising the negative impact that impedes the smooth implementation of the concept of joint management of forests. Removal of exploitative stipulations and greater delegation of power to the lower ranking forest officials and even to the people to the extent possible could form an important part of this entire process. The practical implementation of the latter would greatly lighten the burden of the forest officials who find themselves torn between these contradictory objectives. Since much depends on the initiatives and creativity of the Range Office functionaries, a certain freedom of action would seem to be a necessary precondition for effective management and protection of the forests.

The same applies to the FPCs. Having received recognition from the Government, they are subjected to rules that curtail their independence. For instance, forest officials are expected to be present at FPC meetings particularly when certain decisions are to be taken. It is a standard rule that every offence be reported to the Range Office. Such rules certainly curtail the authority of the committees. A certain degree of flexibility may be discreetly woven into the rules and regulations for promoting effective, and efficient joint management of the forests. In order to concretise the concept of people's participation further decentralisation of management needs to be built into the Forest Policy itself.

Although already alluded to earlier, the role of voluntary organisations needs to be further emphasised. In view of the observations made and given the practical constraints of the Forest Department in promoting and implementing the FPC movement, it seems imperative to seek greater involvement of voluntary organisations. Some officials admitted that this was a task beyond the capacity of the Forest Department's resource constraints. There are also gaps in the training and skills of some of the lower ranking forest officials such as the Beat Officers, guards, etc. Being a relatively specialised task, these officials felt that the task of educating, motivating and orienting the people towards forest protection should be entrusted to specialists like the NGOs. While this was emphasised by an officer whose task involved more interaction with the people, a senior official contradicted this viewpoint and staunchly held that there was no need for the NGOs to intervene, since the Forest Department was competent enough and possessed the necessary infrastructure to cope with this problem.

Field evidence, however, tends to support the former viewpoint. Each Range Officer is in charge of a vast area ranging from 50 to

95 square kilometres. Each Range has about five to seven Beats and each Beat is around 1,000 to 1,500 hectares. Each Beat Officer has on an average two or three guards and two or three watchmen. The number of villages within a Range could vary from 250 to 350. Given this vast area and a limited staff for a task that is of a predominantly administrative nature, it seems a rather formidable job for the Ranger and his officers to simultaneously undertake the task of motivating and orienting the people towards forest protection. If, on the other hand, this task were to be entrusted to the NGOs wherever they exist, the joint management of the forests would be greatly facilitated. The case of Balia is an extreme but very pertinent instance of this which, as was noted earlier, functioned optimally as long as the RDA was motivating and supporting the people but the moment the RDA withdrew, the motivation of the villagers began to wane. While it may be argued that Government functionaries could have been the victims of the same set of circumstances and that it is the circumstances and not the implementing agency that is the root cause of malfunction, it is still highly significant that a voluntary organisation could venture into a task in a highly explosive area and achieve considerable success within a limited period whereas the Forest Department did not have the capacity to take such a major step. The role of the NBS in villages like Korapara, etc., is also a strong case in point. The contribution, therefore, that an NGO can make as an intervening agency in motivating the people seems very significant. It is, in fact, in recognition of its incapacity in undertaking the task of motivation on a large scale and the relative ease with which NGOs can do it that the Government of West Bengal collaborated with voluntary organisations at the State level like the Institute for Bio-Social Research and Development (IBRAD), Calcutta, for the FPC programme. However, the task of the IBRAD is limited to training the government officers and other functionaries regarding people's organisation. The Institute has been quite successful in its task of injecting this new ideology into the general work of the functionaries. However, it must be noted that this input in itself may not be sufficient in view of the statistics presented above, as well as the fact that the FPC movement has spread far beyond its original proportions and has now reached areas which are ridden with topographical and ideological problems. It is in these new areas that the NGOs can play the even more important role of

catalysts. An officially recognised collaboration with the local NGOs via organisations like the IBRAD can be a good solution to the existing problems—especially in areas such as Ranibandh and Khatra forest ranges.

REFERENCES

Bandhyopadhyaya, J. and V. Shiva (1987) 'Chipko'. *Seminar*, February, pp. 33–39.

CSE (1985) *The state of India's environment: A second citizen's report.* New Delhi: Centre for Science and Environment.

Chandra, S.N. and Mark Poffenberger (1989) 'Community forest management in West Bengal: FPC case studies.' In K.C. Malhotra and Mark Poffenberger (Eds.), *Forest regeneration through community protection.* Calcutta: West Bengal Forest Department, pp. 22–47.

Datta, P.K. (1988) *Land reforms administration in West Bengal.* Delhi: Daya Publishing House.

Fernandes, Walter and Geeta Menon (1988) *Tribal women and forest-dweller economy: Deforestation exploitation and status change.* New Delhi: Indian Social Institute.

Fernandes, Walter et al. (1985) *Forests, environment and tribal economy in Chattisgarh.* New Delhi: Indian Social Institute.

——(1988) *Forests, environment and tribal economy: Deforestation, impoverishment and marginalisation in Orissa.* New Delhi: Indian Social Institute.

Gadgil, Madhav and K.C. Malhotra (1980) *Report of the field study conducted on behalf of the Committee on Legislative Measures and Administrative Machinery for Environmental Protection.* Bangalare: Indian Institute of Science, and Calcutta: Indian Statistical Institute (Mimeo).

Jodha, N.S. (1990) 'Rural common property resources: Contribution and crisis.' Foundation Day Lecture, 16 May 1990. Society for Promotion of Wastelands Development.

Kulkarni, Sharad (1982) 'Encroachment on forests: Government versus people.' *Economic and Political Weekly,* 16 January, pp. 55–59.

——(1989) 'Towards a social forest policy.' In B. Chaudhuri and A.K. Maiti (Eds.), *Forest and forest development in India.* New Delhi: Inter-India Publication, pp. 139–69.

Malhotra, K.C. and Mark Poffenberger (Eds.), (1989) *Forest regeneration through community protection: The West Bengal experience.* Proceedings of the Working Group Meeting on Forest Protection Committees. Calcutta, 21–22 June. Institute for Bio-Social Research and Development, Ramakrishna Mission, Ford Foundation.

Menon, Geeta (1989) *Impact of migration on tribal women's work and status in Orissa.* Amsterdam: IDPAD (Mimeo).

Muthiah, S. et al. (1987) *A social and economic atlas of India.* Delhi: Oxford University Press.

Planning Commission (1990) *Towards social transformation: Approach to the Eighth Five Year Plan 1990–95*. Meeting of the National Development Council, 18–19 June. Government of India (Draft).

Roy Burman, B.K. (1989) 'Development of forestry in harmony with the interests of the tribals.' In B.Choudhuri and A.K. Maiti (Eds.) *Forest and forest development of India*. New Delhi: Inter-India Publications, pp. 361–80.

Saxena, V.S. (1986) 'Social forestry in tribal development.' In Deshbandhu and R.K. Gong (Eds.) *Social forestry and tribal development*. New Delhi: Indian Environmental Society, pp. 38–51.

Singh, Chhatrapati (1987) 'Law, ecology and forestry.' *Seminar*, 25–27 February.

5 *Legal, Illegal and Socio-Legal Problems of Women in Wasteland Development*

VASUDHA DHAGAMWAR
ENAKSHI GANGULY THUKRAL

Introduction

This paper examines the socio-legal issues that emerge with respect to women's involvement in wasteland development in the context of employment generation for rural women, rural workers' organisations and women's access to land-based resources. The data for this paper were collected through field visits to wasteland development projects and extensive conversations with the staff of several organisations. We looked at the programmes undertaken by the Self-Employed Women's Association (SEWA), the Mahiti–Utthan project, the Vikram Sarabhai Centre for Science and Technology (VIKSAT) and the Aga Khan Rural Support Programme (AKRSP). As a preliminary to field visits, we made visits to all their offices and talked to the concerned persons who were involved with the various programmes. We were informed that the VIKSAT and AKRSP did not have substantial women's participation—if anything, women's presence in their projects was 'incidental'. The women were involved mainly as labourers in activities such as nursery-raising, pit-making, planting, etc. In Mahiti, the approach so far has been to work with the family as a unit. Though there might be a fair amount of women's participation in the programme, it is not a purely women's programme. The selling of the produce, however, is done through the *mahila mandals*. The SEWA alone wholly concentrates on women. Thus, from the point of view of women's involvement in afforestation, the SEWA projects merited the most attention.

We must add however that we received valuable insights from the Mahiti programme and from Anil Shah of AKRSP and G. Raju of VIKSAT. The story of VIKSAT's efforts to form tree growers' cooperatives for afforestation on forest lands was extremely revealing. We shall return to it later.

We visited four villages in which SEWA is working. They were Kokta in Viramgam *taluka*, Baldana and Metaal in Dholka *taluka* of Ahmedabad district and Ganeshpura in Mehsana district. Kokta and Ganeshpura projects are on *panchayat* lands. The others (as well as Dumali in Dholka *taluka*) are on government revenue lands. We had extensive conversations with Renana Jhabvala, the Secretary of the SEWA and Vandanaben and Bharatiben, the SEWA's field organisers. This study is based on meetings with the women themselves, and, of course, on our own perceptions. This paper examines the SEWA's afforestation programme in depth and also other programmes before commenting on the socio-legal problems of wasteland development.

The objectives of the study were to study the impact of wasteland development on the status of women as well as to find out what the best model of organisation for wasteland development. The best model of organisation for wasteland development will depend largely on what one envisages as the objective of wasteland development. Producing good tree cover is not necessarily to be equated with meeting fuel and fodder needs. Generating income or raising the status of women are again distinct goals. But as one sets the goal of wasteland development as including or meeting these objectives, then obviously the form of organisation adopted should be suitable for achieving them all.

One other way of approaching the question is to ask which type of land—that is land owned by which entity—is most suited to wasteland development. These types can be briefly categorised as private land, *panchayat* land, forest land and government revenue wasteland. We have looked at all of them though the extent of our interest has depended upon the women's component in the project. Women's participation as wage labour in overwhelming numbers does not mean their involvement. It is the part they play in decisions about what to plant, how much to sell and what to do with the proceeds that is vital to determining their role in a project and, therefore, this explains our interest in that venture.

LAND AND LAWS

Private Land

Social forestry programmes undertaken on private lands mainly promote farm forestry. They encourage farmers to plant trees on their own land by distributing free or subsidised saplings (CSE, 1985: 52). Usually the farmer plants trees side by side with crops or vegetables. It is his decision as to what species of trees he wants to plant and he requires no one's permission to grow trees, anymore than he does to grow crops. Technical advice, funds and saplings are all he needs. Mahiti and VIKSAT are organisations that have helped to initiate this kind of programme. Funding for this kind of programme is available from various governmental and semi-governmental agencies like the District Rural Development Agency (DRDA), and the National Wasteland Development Board (NWDB) and governmental schemes like the Integrated Rural Development Programme (IRDP), the Rural Labour Employment Guarantee Programme (RLEGP), etc. Banks, too, can be approached for loans.

Panchayat Land

Afforestation schemes on *panchayat* land can be undertaken only after the *panchayat* gives its permission. In most cases, the *panchayat* takes a share of the profit from the produce. The *panchayat* does not lease out or allot the land to the persons engaged in afforestation. It merely gives usufruct rights to carry out the programme. SEWA's programmes in Ganeshpura and Kokta, which are described later, are being undertaken on *panchayat* land. As in the case of afforestation on private land, grants for carrying out this kind of programme are available from governmental and semi-governmental agencies.

Forest Land

In 1982, the Forest Department of the Government of Gujarat passed a resolution (G.R. dated 16 June 1982): 'to give on lease the wastelands, deforested hills and land on which there are no plans for afforestation in the next twenty years (all this land being under the Gujarat Forest Department) to responsible private bodies such as educational institutions, registered societies and public trusts and industrial organisations.' The G.R. laid down that the lease would be for 30 years.

In 1984, another resolution was passed (G.R. dated 27 September 1984) which partially replaced the G.R. of 1982. The latter specified that forest wastelands would be leased only to registered cooperatives of landless labourers and marginal farmers who made a living from the forest land and minor forest produce. The other conditions laid down in the G.R. of 1982 remained unchanged.

Government Revenue Land

The Government of Gujarat issued a G.R. on 1 January 1987 making it possible to do afforestation on government revenue lands.[1] The introductory paragraph to the G.R. states that the government had taken into account nine land-related resolutions (passed between 1965 and 1984) dealing with the national policy for planting fruit and other trees and for afforestation to overcome environmental degradation caused by illegal felling and reduction in rainfall. In pursuance of the national policy, the Government of Gujarat had also passed a resolution to bring 50,000 hectares of land under plantation. One of the objectives of that G.R. had been to create employment opportunities for the rural poor. This G.R. was also taken into account as was the National Wasteland Development Scheme.

The G.R. of 1987 states that in view of the situation in the country as well as the national policy relating to it, the Gujarat Government had decided to bring wastelands and fallow lands under this programme and to plant fruit and other trees on them.

[1] G.R. dated 1 January 1987 (GMN 3986—2226 A). We are indebted to Rani Advani, Advocate, for the translation from Gujarati.

According to the G.R. of 1 January 1987: (*a*) government fallow land, (*b*) government wastelands, (*c*) government sandy lands and (*d*) government saline lands would be brought under afforestation. Government lands already converted to revenue lands, however, would not be brought under the G.R. *Panchayat gauchar* lands were also exempted.

The eligible beneficiaries or the recipients of land under this G.R. were detailed as follows:

1. Gujarat Gram Vikas Nigam (Gujarat Village Development Authority).
2. Tree growers' cooperatives formed by:
(a) Scheduled Castes (SC) and Scheduled Tribes (ST),
(b) landless persons,
(c) marginal farmers,
(d) small peasants,
(e) Other Backward Classes identified by the Baxi Commission.
The tree growers' cooperatives would have to be registered under the Cooperative Societies Act, 1962. Furthermore, all members of the tree growers' cooperatives had to belong to one of the above five categories and had to be eligible for the IRDP scheme.
3. Individual applicants in the following order of priority:
(a) agricultural labourers,
(b) landless labourers,
(c) marginal farmers,
(d) small farmers,
(e) local voluntary agencies.
Although it is difficult to see how the last category can be called individual, even amongst these categories priority would be given to SC/ST individuals or their local voluntary agencies.
4. Industrial enterprises of the government, or enterprises with substantial government interest or joint ownership with government.

The G.R. also specified that the land was to be given on lease with the ownership remaining with the government. The quantum of land to be leased and the conditions on which it was to be leased were also laid down. The Gram Vikas Nigam or voluntary agency or tree growers' cooperative would be given a maximum of two hectares of land per member. No Gram Vikas Nigam could hold more than 80 hectares. In other words, the membership could be between 40 and 80 members. Individuals would similarly get one or two hectares of land.

The land was to be given on a 15-year lease. No rent was payable for the first eight years after possession was given. Thereafter, half the rateable value assessed from the surrounding agricultural land would be payable as rent. After 15 years, the lease could be extended for one year at a time on full agricultural rent determined by the above-mentioned method.

Every lease-holder was required to plant 1000 trees per hectare. These could be fruit or other trees. If some land still remained available thereafter, the lessee could grow crops of grass on it for personal use. In special circumstances, the Collector could, after consultation with the forest and environment departments, reduce the number of trees required to be planted on one hectare. The lease was subject to right of way or easement. An authorised officer could order clearing of the land in public interest. If the lessee had spent money on improving the land, the Collector had discretionary powers to award compensation.

The land under lease must be developed within *three* years, failing which the lease would be cancelled. However, if sufficient and genuine reasons were shown for failure to do so, the period could be extended by a further two years. If the lease-holder still failed to develop the land then, at the end of the extended period, the lease would be cancelled without compensation. If the conditions of the lease were violated and the land was put to other uses the lease would be cancelled. The lease was heritable and the leased land could be given as security for getting a loan from a recognised bank for developing it. The government and related enterprises were required to develop the land in such a way as to generate employment.

A fairly detailed procedure for deciding who would get leases was also laid down. The committee set up to do so included the *prant adhikari*, the Director of the Gram Vikas Agency, the *mamlatdar* or *tehsildar*, and the State Social Welfare Officer. The committee would be required to have a detailed list of all land available in the *taluka*. It would examine the applications and decide on them. Applications of industrial enterprises had to go through a lengthier procedure and had to satisfy the Collector, the Revenue Department, Department of Rural Development, Social

Welfare Department and Adivasi Development Department. Thereafter, the Government would take the decision on the applications.

Afforestation on Panchayat Land: Ganeshpura and Kokta

Ganeshpura

The oldest wasteland development project of SEWA is Ganeshpura village in Mehsana district. It was set up in 1985. The idea of taking up forestry was the result of SEWA's promotional activities for smokeless *chulhas*. During the time that the smokeless *chulhas* were being promoted, the local women felt the acute shortage of fuel and fodder. There was a felt need to grow trees on wastelands for overcoming this problem. Thus the initiative for this programme came from the women themselves.

As part of the wasteland development efforts, four training programmes were organised, mainly with the cooperation of the local Forest Department. These programmes dealt with the following themes:

Rural women and forestry;
How to make a nursery;
Water management and land development by poor women;
Managing the cooperatives.

Some of the women who undertook nursery work started planting trees on their own lands. From the nurseries emerged the idea of 'plantations'. At this the landless women expressed a desire to undertake a similar kind of activity, and persuaded the *panchayat* to give them usufruct of 18 acres of *panchayat* land. In an agreement between SEWA and the *panchayat*, it was decided that after three years when the produce could begin to be sold, the *panchayat* would be given a third share of the profits.

In 1985, the women began work on ten hectares of land.

Funding from NWDB enabled SEWA to pay the women wages for preparing the land and digging the pits. Twenty women became involved in the programme. However, 1985–87 proved to be drought years and the mortality rate of the saplings was very high. To overcome the problem of water shortage, SEWA obtained a grant from the Council for Advancement of People's Action and Rural Technology (CAPART) to dig and line a pond. Another grant from the CAPART also enabled them to sink a bore on their land.

When we visited Ganeshpura we found a thriving ten acre plantation covered with *neem*, tamarind, *prosopis juliflora, pendula* and *israeli baval*. The women had also planted vegetables and fruit trees in between the trees. Some income was already being generated from the latter. Grass was already available for fodder. The women were planning to extend their activities to five more acres where they intended to plant trees and mustard seed (an oil crop). We asked the women to tell us about who decided which trees to plant. They replied that the decisions were made with the advice of local farmers and the Forest Department.

The women's perceptions about the afforestation project were very clear. It was a source of income, and it met their fuel and fodder needs. The vegetables and fruit were additions to their income as well as to their diet. Most important, the women had a sense of security and continuity in doing their own work; they said they were no longer at the beck and call of any employer and this was very important to them. We were not sure how far they had understood the ecological value of afforestation though they had been told about it. They informed us that they would fell trees such as *neem* trees after 15 years, sell them for timber and plant fresh trees.

The Ganeshpura project faced two problems—one was with the Rabharis or cattle-herders who had been grazing their cattle on that land. This problem was sorted out by appointing one of the Rabhari men, who was respected in his community and in the village, to be the watchman.

The second problem is of more recent origin and goes deeper. The Ganeshpura programme of wasteland development began as a SEWA project. For some time SEWA has been trying to set up the Vanalakshmi Mahila Mandali in Ganeshpura as an independent cooperative, but has not been successful because of conflict between two communities, to wit, the Prajapatis and the Harijans. This conflict has existed from the very beginning, but the SEWA organiser, Bharatiben, told us that they had somehow coped with it. Now at the time of registration of the cooperative this problem had cropped up once again. To begin with, 45 women had agreed to become members, of which the 15 Harijan women tried to back out. However, further investigation revealed that the problem was actually being created by a Harijan (Senma) labour contractor who had been providing labour to the Forest Department. His fears were that once the cooperative was registered it would divert much of the labour to itself and would also start contracting labour to the Forest Department, thereby cutting into his market. Renana Jhabvala mentioned that it was possible that the contractor had a fake cooperative and feared that if a real one came up, it would cut into his scheme. He had managed to convince the Harijan (Senma) women to side with him. What perhaps did not help was that the SEWA organiser of this village is a 'Prajapati' woman.

Kokta Village

In village, the initiative for wasteland development came from the *panchayat*. The *sarpanch* of Kokta told us that at a meeting of *sarpanches* the 'Government' told them to undertake wasteland development. He was very impressed by what they had been told about the benefits of afforestation and even quoted a verse which said that one tree is equal to ten sons. He had heard of SEWA and approached them to take up the work. The *panchayat* was willing to give them two acres and 18 *gunthas* of land on the condition that three years from the inception of the project a third of the profits from the sale of the produce would go to the *panchayat*. For the first three years no money would be due.

The SEWA is helping to set up a demonstration plantation on this plot. The land had earlier been set aside by the *panchayat* for the *anganwadi* programme, which was supposed to become self-sufficient by taking up farming or poultry-rearing on it. It had been lying uncultivated, however, for a very long time.

The programme is very recent. This is the first year for planting.

During our visit eleven women were involved in clearing the land and planting saplings. Other women were also busy preparing to plant vegetables such as tomatoes and eggplants (brinjals).

The women seemed to feel that this project would be a source of income, though one of them did voice the opinion that outside wages were higher than the Rs. 16 she earned here. They seemed more excited about the possibility of getting vegetables to sell and eat in the same season than about the produce from the trees which would obviously mature several years later. The availability of fuel and fodder was also attractive to them. Even more significant was the women's involvement with the programme and their sense that once this project got underway they would not be dependent on outside work again in the same way. When we asked them about their plans for the income generated by sale of the produce, the women said that one-third would go to the *panchayat*, a third would be divided amongst them and the remaining one-third would go into their savings account.

Afforestation on Revenue Wasteland: Baldana and Metaal

SEWA has been working in several villages in Ahmedabad district since 1977. When the minimum wage for agricultural labourers was declared, some of these agricultural labourers came to SEWA for help because the area was dry. Employment in the area was available for about four months of the year, and the rest of the time the labourers had to migrate as far as Surat and Saurashtra in search of work. Even to be able to get the minimum wages, work was required. Since the area was dry SEWA decided on starting a dairy and the making of handicrafts.

Having discovered that many of the agricultural labourers were weavers who had put away their looms, SEWA decided to train the women of Baldana and Metaal—who had traditionally never sat at a loom—to weave. Besides this, other crafts like making pottery, files, *khadi* and baskets were also undertaken. The major activity was the women's milk cooperative. These activities were also extended to Baldana and Metaal villages which we visited.

During the drought of 1985 the women had faced a problem of shortage of fodder for the cattle. That is when the idea of waste-land development and afforestation to meet the fodder and fuel needs was brought up. But by the time the programme could actually commence, it was well into 1987 because the whole process of getting funding, land on lease, etc., took time. Renana Jhabvala informed us that one of the reasons why it took so long to get the land lease was because most government lands over 50 hectares in Gujarat were being kept aside to resettle the Narmada oustees. Allotment of large plots of land therefore proved very difficult.

Initially SEWA had planned to start the afforestation work as a SEWA programme and then to register the village cooperative societies gradually and allow them to be separated once the programme took off. However, as Jhabvala told us, SEWA was compelled to start with registering the village cooperatives first because as per the G.R. of 1987, the land could not be leased directly to SEWA. As we have seen earlier, the G.R. enumerates the possible beneficiaries of the government wasteland afforestation scheme. An NGO like SEWA does not fall into any of the categories. Therefore, three separate cooperatives were registered in 1987 in Baldana, Metaal and Dumali villages. They were all allotted land in various quantities on 6 May 1988.

Baldana Village

In Baldana where SEWA had already been running a training and weaving centre, the cooperative was allotted 185 acres of land on lease. Work in Baldana began on a small plot in the SEWA centre on a demonstration basis. Various tree species including *neem*, tamarind and subabul were planted. Vegetables were also planted between the trees. But 1988 proved to be a bad year due to drought, so the saplings did not survive and other had to be planted. During our visit in 1990 we saw that the trees had grown during the last two and a half years and the women have also been able to grow some vegetables. In 1990, they planned to experiment with other plant species, like banana, mustard, etc. The demonstration centre is living proof of the good things afforestation can bring.

The land leased to the cooperative is a little distance away from the main village. A stream which had waist-deep water when we visited the spot in October 1990 had to be crossed to reach the land. Though work had begun on it earlier in 1988, it was unsuccessful till 1989 due to drought, floods and the hostility of the Bharwads who are cattle-grazers. The Bharwads showed their displeasure by forcibly grazing their cattle and uprooting the saplings in the afforested land which they regarded as their traditional pastures. This problem has now been settled (we shall return to it later). The year 1990 witnessed bunding, fencing of plots and planting of saplings of tamarind (*khatta imli*), subabul, *ganda baval*, *neem*, *desi baval*, *ram baval*, *israeli baval*, *seras* and *mehendi*. The women we met were hopeful that the plants would survive. For this work they had received a wage of 50 paise for digging one pit and 30 paise for planting one sapling. They planned to grow vegetables in between the saplings.

The women definitely perceived this work as their 'own' and had a sense of pride in it. The fact that they did not have to constantly migrate to construction sites and textile mills for jobs as they used to do earlier seemed to give them a sense of security. They also saw this activity as gainful employment, as one which would also help them meet their fuel and fodder needs and improve their nutrition. Income from selling vegetables was also part of their calculations. But they said that even if they earned less on their land than outside they would still remain on their land. The sense of continuity was most precious of all to them. Indeed, they told us that they planned to stay on their plots in order to look after their plants in a better way. These women were all landless. Each woman had been allotted two hectares of land for 15 years. The sense of security the 15-year lease had given them was remarkable. We asked the women to tell us about those who decided what trees should be planted. They replied that the Gujarat Agricultural University had been consulted in the matter. When asked what they would do with the vegetable produce they said that they would eat them as well as sell them.

Metaal Village

A total of 219 acres of government revenue wasteland was

allotted to the Metaal women's tree growers' cooperative—Shri Vanraji Mahila SEWA Vruksha Utpadak Sahkari Mandali Limited. As in Baldana, the SEWA began with a programme for training women in weaving. But the programme in Metaal did not take off. Some of the SEWA members were then sent for nursery training. In fact, the saplings that the women were planting in Baldana were from the Metaal nursery.

The land was prepared and some of the saplings were planted in 1988 when the problem arose. The Bharwads, as in Baldana, started uprooting the saplings or letting the cattle into the plots to destroy the saplings. Prior to being allotted to the cooperative, the Bharwads had been grazing their cattle on this land. Even though they knew the land did not actually belong to them, they saw themselves as having usufructuary rights over this land, and saw the allotment of the land to the women as a threat to their 'traditional' rights. The *panchayat* supported them. Finally they took legal action. Thus began a long drawn out legal wrangle between the Bharwads and the *panchayat* on the one hand and the cooperative on the other. (This has been dealt with in greater detail later in the paper.)

Even after a compromise was worked out, the Bharwads continued to destroy the saplings. Finally, Vandanaben, Assistant Project Coordinator, said that she would not help the cooperative to resume work until the *panchayat* gave a written undertaking that it would bear responsibility for any deliberate damage to the saplings and for maintaining law and order. Although the women were keen that the programme should start, they were unwilling to take responsibility for the plantation. Tulsibhai, who is the full-time SEWA worker stationed at Baldana, felt compelled to tell the women in Metaal: 'The women in Baldana are bold, and willing to take responsibility for their land and plants. They too have given away 15 acres to the Bharwads and they are now brave enough to face up to them and tell them off if they let their cattle into the women's lands. Are you willing to do the same? Otherwise you can forget about working this land.'

Initially, the *panchayat* was quite supportive, but in the meantime, elections were held and the *sarpanch* changed sides. This had also happened in Baldana.

We did not visit Dumali. But from the information we received from talking to Vandanaben and Tulsibhai, it seems the women in

Dumali, apart from the droughts, faced the least problems. As a result, the programme was running fairly smoothly. The biggest problem they had faced was when the Padhers (a community of earth-cutters) from the neighbouring village of Shiyal had attempted to lay claims on the land. The *panchayat* had driven them away and told them that the boundary of their wasteland lay outside the Dumali village boundaries and therefore they were not to enter the Dumali wasteland.

We will deal briefly with the activities of the three other organisations we visited: Mahiti, VIKSAT and AKRSP. Though their programmes are not women oriented, their experiences are extremely useful in the context of any study of social forestry.

Mahiti–Utthan

Mahiti–Utthan, popularly referred to as Mahiti, is a voluntary organisation based in Ahmedabad and is working in the villages of Dhandhuka *taluka* in Ahmedabad. According to Nafisa Barot, the Executive Trustee of Mahiti, there are 72,000 hectares of wasteland in the Bhal area of Dhandhuka *taluka*. The land is totally saline and completely unfit for cultivation. It is also devoid of water. Every year the people and the cattle are forced to migrate in search of water. In addition, the people also migrate in search of work.

Mahiti is at present working in eight villages. We were informed by the members of Mahiti that among their different activities, the most important development activity of the organisation is social forestry. Related to this activity is their programme of water management. As part of this programme, Mahiti helped dig nine water tanks in the eight villages and lined them so as to prevent salination and seepage. This helped to tackle the severe problems of scarcity of drinking water for cattle and humans.

In this social forestry programme, Mahiti has also initiated the planting of *piludi* trees on the private lands belonging to the farmers. This work is being carried out with a grant from the NWDB. They have also helped 54 landless persons to obtain land under the Gujarat Government's tree patta' scheme under the G.R. of 1987. Mahiti is also helping the villagers to get bank loans for planting purposes.

Some villagers are also growing *piludi* trees on revenue wasteland. This land has been allotted to Mahiti which sublets it to the villagers. The villagers could not take the land on lease directly as they had not formed a cooperative. Members of Mahiti told us that the villagers were not equipped to handle the programme on their own and hence registration of the cooperative had been postponed. Once the plantation takes off, Mahiti plans to help them form cooperatives and will then 'transfer' the leases to the cooperatives.

As mentioned earlier, we did not look at Mahiti's social forestry programme closely because Mahiti is working with the *family* as a unit. Though women are involved, it is not a programme designed specifically for them. The marketing of the *pilu* fruit, however, is done by the *mahila mandals*.

PROBLEMS: LEGAL, ILLEGAL AND SOCIO-LEGAL

Three out of the five projects that we discussed with SEWA (we visited four) were on government revenue lands, and two were on *panchayat* land.

As one can see, the G.R. of 1987, which deals with afforestation of government revenue lands, is quite detailed in its requirements. The conditions are also somewhat stringent. The land is given only on lease. While 15 years is a reasonable stretch of time (though forest land is given for 30 years), it is reduced drastically to three and at the most five years in which to show a successful development to the government. What would constitute a success? How is development to be defined or measured? What is the expected survival rate of plants? What would constitute sufficient and genuine cause for not developing the land? What is the public interest for whose sake a plantation may be destroyed? While excessively precise instructions or conditions create problems, vague or overly general instructions can also create legitimate unease and may leave too much to official discretion. However, we must add that the three year continuous drought which had resulted in almost total failure of the plantations had not led to cancellation of SEWA's leases. Nevertheless, the possibility of the lease being terminated or plantation being cut down does remain and can lead to feelings of insecurity as also the fact that after 15 years the lease is renewable only for one year at a time. The last condition is not designed to permit any long-term planning after the initial lease is over. Given that the fruit and other trees are included as part of the definition of social forestry and not conservation forestry, the trees will have to be replaced at regular intervals. This is because they are to satisfy the fuel, fodder and timber needs, or, in the case of some trees, because they complete their lifespan. In other words, these plantations will not regenerate themselves without some long-term planning and attention. Short leases will not satisfy this requirement. But these problems are in the future, at least for SEWA. The major problem that can arise: if the next two years do not produce the expected level of land development in the leased lands is that then the Collector shall cancel the lease. While the Government clearly needs to assert its ownership of the land and is not willing to jeopardise it with long leases, short leases, as we have seen, could be counter-productive.

Dumali, Baldana and Metaal all have received leases of government revenue wasteland. All three tree growers' cooperatives were registered in 1987 with 20 members each. Kokta and Ganeshpura projects are on *panchayat* land. In neither village has a tree growers' cooperative been formed by the women so far. In four villages (Dumali, Metaal, Baldana and Ganeshpura), the women's groups have faced and are still facing problems. While the law (as an instrument of harassment, intimidation and delay) has been put to good use, one cannot say that the problems are legitimate or purely legal problems. At best they are socio-legal problems.

All three village cooperatives obtained their land leases on 6 May 1988. At that time Assistant Collector who made the lease stipulated that all the cooperatives should increase their strength from twenty to the full complement warranted by the size of the leased land within three months. It should be remembered that as per the G.R. of 1987,[2] while each member is entitled to one to two hectares of land the maximum strength of the tree growers' cooperative can be 80 members.

In Dumali, Metaal and Baldana villages there have been disputes over the leased land between the women and other residents who laid claims to the land. In Dumali village, the Assistant Collector Dholka allotted 208 acres and 24 *gunthas* of land to the cooperative. However, the Padhers of Shiyal village laid claims to the land. The Padhers were outsiders and the *panchayat* took up the fight on behalf of the women. The Padhers were shown effectively that the land in question was in Dumali village and the Padhers could have no claim to it. The matter did not go to court and was sorted out locally. The *panchayat's* support to the women as well as the Padher's status as outsiders were crucial to the resolution of the dispute.

Baldana Village

In Baldana, 185 acres of land were allotted to the women's tree growers' cooperative. Here the disputants were Bharwads, a cattle-herding community. They were also residents of Baldana. For two years, the Bharwads thwarted the efforts of the women to plant saplings by letting their cattle into the plots. The Bharwads' chief grievance was that after the lease was given out they no longer had land on which to 'rest' their cattle. They wanted land that was higher than the rest, so that rain water could drain away and allow the cattle to be comfortable and dry. The matter was settled by the *panchayat* and the women agreed to give up some land to the Bharwads. They were not very happy about it, but they could see they had little option. This was both in their short-term and long-term interest. The Bharwads had kept disrupting their work for two years. Besides, they all had to live in the same village and good relations were of paramount importance. The women gave up around 10–15 acres of land to the Bharwads (they were not sure of the exact amount). Vandanaben said that it was prime land. The women had dug a small tank on the land to store water for the plants. The Bharwads were allowed to have the land on the condition that their cattle would not drink from the water tank. The women put fencing round it. As Vandanaben informed us, it was not at all certain that the Bharwads would honour this

[2] G.R. of 1 January 1987 (IV Conditions of Lease, Para. 4.6) states that all leases are subject to right of way.

condition. The agreement between the women's cooperative and the Bharwads was not a written agreement. However, since then the Bharwads have not trespassed on the remaining land. According to Tulsibhai, the Baldana women have brooked no nonsense from the Bharwads. During the 1990 monsoon season, the women once again took up work on the land: bunding, fencing with cactus, and planting saplings. In addition to the land given up to the Bharwads, the women surrendered five to seven acres of land for a road, and they put aside ten acres for demonstration forestry. Thus out of 185 acres they have about 160 acres of land on which to plant trees. Surrender of land for road building is entirely within the terms and conditions of the leases.[3] It can be argued that the demonstration plot is necessary for proper development of the remaining land. But a problem could arise over the agreement between the women and the Bharwads.

It should be noted that the land does not belong to the women. They cannot alienate it. They cannot legally transfer it to the Bharwads. They have no authority to give anyone else the licence to use that land. The Government of Gujarat is not a party to this disputed resolution. If there was a written agreement, the Government could use it to prove violation of the terms and conditions of the 15-year lease. On the other hand, as the agreement is unwritten, (and given the power of the written word in rural India) the verbal agreement might not be honoured as strictly as the written one. Vandanaben herself was not too sure about the inviolability of the water tank. Furthermore, as the terms of the lease require that the land should be allocated to the members in such a way that no one has more than two hectares of land, if the agreement with the Bharwads has no validity in the eyes of the Government, there would have to be two more extra members of the cooperative to whom the land actually developed would have to be apportioned. Fifteen acres is a small area but this may not always be the case, as indeed has happened in Metaal village.

SEWA admittedly is an organisation which has a great deal of status and prestige. A tree growers' cooperative under its umbrella may not have such problems with the Government, but

[3] G.R. of 1 January 1987 (IV, Conditions of Lease, Para. 4.6) states that all leases are subject to right of way.

other agencies would most likely have difficulty in a similar situation.

However, as a positive result of the agreement, work has begun again on the Baldana lands in the third year of the lease. The women have faith in the agreement and informed us that the Bharwads could not now enter this land—the *panchayat* would not stand for it. They themselves had planted a cactus fence round each two-hectares plot and were even capable of shouting at any cattle-grazer whose cattle were likely to trespass.

Metaal Village

Unlike Dumali and Baldana, the Metaal village cooperative has not been able to sort out its problems. The conflict turned into a protracted legal battle and is an excellent illustration of how, in skilled hands, law can be turned into an instrument of harassment, delay and intimidation even when the harassed party has all the legal rights on their side.

In Metaal, the tree growers' cooperative had been given a lease of just over 219 acres of land. As with the other two villages, the Metaal cooperative also obtained the lease on 6 May 1988. On 16 May 1988, Manju Raju, SEWA Project Coordinator, wrote to the Revenue Secretary that the Assistant Collector of Dholka *taluka* had passed an order that the cooperative should increase its membership to the full number, (warranted by the size of the land) within *three months*. She reminded the Revenue Secretary that, in a discussion between SEWA and the Revenue Secretary, it had been decided that the work of developing the land would be phased out and membership increased accordingly over a period of two to three years. Looking at the terms and conditions of the lease as per the G.R. (paras. 3.2 and 3.3) the Assistant Collector was within his rights to make such a demand. On the other hand, this condition is not practicable from the point of planning and organisation.

On *16 June 1988* the cooperative got possession of the land. On *8 July 1988* the Assistant Collector amended his orders and extended the period for reaching full membership to two years. One wonders how far an ordinary village cooperative could have succeeded—if it could get round the paperwork at all—in dealing with officialdom, especially when the latter is only applying the strict letter of the law.

Members of the Metaal women's cooperative, the Vanraji Mahila SEWA Vruksha Mandali, told us that saplings were planted during the rainy season of 1988. But the Bharwads drove their cattle onto the land and the saplings were either trampled or eaten. Thereafter, the Bharwads and the *panchayat* took recourse to the law.

In *January 1989,* the Metaal *gram panchayat*, the *sarpanch* of Metaal *gram panchayat* and one resident of the village filed a Civil Suit (No. 48/49) in the Court of the Civil Judge (Senior Division), Ahmedabad, against the Government of Gujarat and Metaal women's cooperative. The prayer was for a permanent stay against the lease of the 219 acres to the women's cooperative on the following grounds:

1. The land allotted to the women's cooperative was *panchayat* wasteland and *gauchar* land.
2. 2,500 heads of cattle grazed on that land and there was no alternative pasture land for them. Also there was no authority for using the land for any purpose other than grazing.
3. The land had been allotted to the cooperative without giving the plaintiffs notice of that intention or hearing their objections.
4. The Assistant Collector had wrongly refused to give them a copy of his orders on the ground that they were not persons interested in the matter as their interests were not affected. But their interests *were* affected.
5. Acting in a judicial capacity, the Assistant Collector was duty bound to give them a copy of the order.
6. The proceedings were vitiated because they were against the principles of natural justice and the plaintiffs were asking for a stay order so that their cattle-grazing rights may not be affected.

The defendants, i.e., the tree growers' cooperative (but in fact SEWA) replied to each of these contentions in *March 1989* and denied all the allegations. The land was not *panchayat* wasteland or *gauchar* land. There were not 2500 head of cattle. There was no requirement that the government must notify the *panchayat* before allotting land, and in any case this was government land. The Government could allot it to anyone they wished. Further, the plaintiffs had not been notified because they were not interested parties.

On *14 May 1989*, the Civil Judge (SD) rejected the application for a permanent stay.

In *June 1989*, a second attempt was made to do away with the cooperative. Four individuals and the Metaal *gram panchayat* filed a fresh petition, this time before the *Collector* of Ahmedabad District. The petitioners said that they were aggrieved by the allotment of the land in question to the Metaal tree growers' cooperative by the Assistant Collector Dholka on 6 May 1988. They argued that the order of the Assistant Collector was bad in law, irregular in following procedure, without jurisdiction and violative of their rights to graze their cattle as well as their rights of easement. They also invoked the Gujarat Land Ceiling Act and said that the allotment violated the provision of the Act. It also alleged that the *mandali* was a 'bogus' society. The petitioners said that they themselves were not to follow any procedure as they were interested persons. They pleaded that the delay in making this application to the Collector be condoned *as they had only come to know of the Assistant Collector's order in May 1989*.

It may be noted that the *gram panchayat* was party to both the suit before the Civil Judge (SD) and the application before the Collector. The petitioners in the Civil Suit had applied for a certified copy of the order of the Civil Judge (SD) on *22 May 1989* and the same was ready for delivery on *12 June 1989*. The lawyer for the applicants in both matters was the same.

On *25 June 1989*, the Collector passed an *ex-parte* stay order.

On *30 June 1989*, SEWA received telegraphic intimation of the stay order from the applicants's lawyer.

On *3 July 1989*, the Collector amended his order, fixing the date for hearing on 18 July 1989 and giving stay till 18 July. However, the very next day, on 4 July 1989, evidently on the respondent (the cooperative through SEWA) putting forward the facts of the case, the Collector passed fresh orders. In his order the Collector pointed out that the *gram panchayat* was party to both the civil suit and the application before him. The civil suit had been filed in January 1989 and had arisen out of the Assistant Collector's order. Therefore the claim that the applicants were ignorant of the Assistant Collector's order was patently untenable. The lawyer in both matters was the same. The Collector held that the application was barred by limitation and that the applicants did not have sufficient cause to file their application. He vacated the stay and dismissed the application.

The Metaal *gram panchayat* and others in its party had not yet given up hope of getting rid of the Vanraji Mandali.

On *10 July 1989*, a fresh civil suit was filed (C.S. No. 163/89), this time in the Court of the Civil Judge (Junior Division), Ahmedabad district. This time the Metaal *gram panchayat* was not one of the plaintiffs. The lawyer was different. The arguments were similar. It was claimed that this was grazing land (the word '*panchayat*' land was not mentioned). Violation of personal interest and customary rights was argued. It was further added that the *gram panchayat* has passed a resolution to the effect that for every 100 head of cattle, 40 acres of grazing land were needed and, therefore, this land could not be spared.

On *18 July 1989*, the Civil Judge (JD) gave an *ex-parte* stay order till the next date of hearing (not specified in our records). SEWA was informed about the stay by the plaintiff's lawyer.

The police obviously had 'problems' understanding the meaning of 'stay'. So much so that on *29 July 1989*, Manju Raju, the SEWA Project Coordinator, wrote to the Assistant Collector Dholka that the land was already in their possession and they were working on it. 'We now understand that the Police Sub-Inspector has been instructed to take action against us for working on the land.... Could you give instruction to the Police Sub-Inspector explaining the meaning of the status quo ante.'

It is to be noted that each time the plaintiffs failed to get a stay they abandoned that suit or application altogether. They were very clear that a case in which no stay had been obtained was only a waste of time and money for them, while having no harassment value. The Civil Judge (JD) gave a stay, but as the women were already working on the land it did not help the plaintiffs which is why that attempt was also abandoned.

Finally, the whole dispute was sought to be solved by a settlement or *samadhan*. On *21 August 1989*, a settlement was arrived at in the presence of the Collector, Renana Jhabvala (Secretary, SEWA), Manju Raju (Project Coordinator) and Vandanaben (Assistant Project Coordinator). The *sarpanch* of the Metaal *gram panchayat* and other prominent villagers were also signatories to it.

According to this settlement, the *mandali returned 108 acres of land to the Government*. This returned land was to be used solely for grazing. Vandanaben said that 12 acres were additionally set aside for the annual *mela* or fair. (The fair being a short yearly event, this land would naturally be available for grazing for the rest of the year.) Such Bharwad women as could be made members of the cooperative (as per the G.R.'s requirements) would be admitted to membership. In their turn, the village agreed not to litigate against the cooperative. The agreement was being made in order to maintain peace and harmony between the concerned parties. As a result of the agreement, women were left with 80 acres of land (the figures do not add up to 219 acres, but more precise information was not available). It is to be noted that: (*a*) the agreement is headed 'In the Court of the Collector, Ahmedabad District' and thus the Collector was officially informed about the agreement, (*b*) the land was returned to the Government and (*c*) it was earmarked for grazing, a condition that might bind the Government if it tries to put the land to other uses in the future.

The agreement took place in August 1989. In September 1990, there was still no development of the remaining land. This was because the women were not willing to go ahead without an assurance from the *panchayat* that it would take responsibility for deliberate destruction of their plants. The *panchayat* was not willing to do so. It may be recalled that in Baldana the *panchayat* had agreed to support the women after the settlement.

The Use of Law

While the law has been used very skilfully and neatly to defeat all attempts of the *mandali* to develop the truncated portion of the land originally allotted to them, the problems of Metaal, Baldana or Dumali cooperatives are really not legal problems. They arise out of a number of socio-economic and even political realities. We were told that in both Baldana and Metaal the *panchayat* elections were won on the assurance that land would be restored to the Bharwads. Renana Jhabvala told us that, after the Baldana agreement, a very satisfied *sarpanch* exclaimed: 'This means 600 votes for me.' Thereafter the *panchayat* has thrown its weight behind the agreement. However, it is difficult to say what the *panchayat* will do or say if the *mandali's* stored water is used up by the cattle of the Bharwads.

Even though they have successfully wrested at least half the land originally allotted to the women's cooperative the Bharwads of Metaal village are still not content. This is why the *panchayat* is refusing to give their assurance of support to the cooperative. SEWA's field organiser is being told by the Bharwads to go ahead with the plantation work but behind her back the women are threatened. The Bharwads now want all their women to be enrolled as members of the tree growers' cooperative. This is manifestly not possible as the G.R. states that members must come from specific disadvantaged classes and as most Bharwads in Metaal are affluent, they cannot qualify under the G.R. Indeed, we were told that the *sarpanch*, the *talathi*, the Collector, the record-keeper (very powerful by virtue of his control over land records) and the Bharwads gave SEWA a list of 90 women from affluent castes whom they wanted to be made members of the cooperative. Vandanaben replied that those women did not satisfy the rules for enrolment and moreover, there was not enough land left with the cooperative to accommodate another 90 women. The *talathi* then produced a revised list of 30 women, including ten Bharwads. The other 20 were from Vandanaben's list. She would have preferred to take only five, but could not get the Bharwads to agree to it.

The Metaal *panchayat* is supporting the Bharwads and shows every sign of continuing to do so until either the Bharwads reduce their demands or the women give up. In the meanwhile, two of the first three years during which land must be developed are over. One does not know whether the Government will extend the lease beyond May 1991 which it can do only if it sees the reasons for not developing the land as reasonable and sufficient, genuine though they are.

To return to the cases initiated by the Metaal villagers and the *gram panchayat* against the women's *mandalis*, they clearly exemplify the way in which law can be used to intimidate, harass and delay. They also exemplify the extent to which the system ignores such harassment. Thus, in disposing off the application before him the Collector noted that the petitioners had lied in deposing that they had not heard of the Assistant Collector's order allocating land to the tree growers' *mandalis* and that in fact they had filed a suit against it in the Court of the Civil Judge. The Collector also noted that the lawyer in both cases was identical. Yet no action was initiated against the lawyer in question for perjury. He was not even reported to the President of the Bar. Just as important, it was not made clear to the petitioners that they could not thwart the will of the State. The *samadhan* entered into between the women's cooperative and the village in the presence of the Collector is in fact a tacit refusal of the Government to assert itself by backing the women. The village elite may not have won the battle, but they have won half the war, and may be close to winning the other half.

Just after the first suit filed by the Bharwads and the Metaal *gram panchayat* in the Court of the Civil Judge (SD) had been disposed of, the ILO team had discussed the Metaal problem with SEWA. We quote:

> The filing of the case caused consternation amongst the women, but the SEWA organisers regarded it as another open challenge to the attempt of poor women to get rights to land. Ms Ela R. Bhatt, General Secretary of SEWA, expressed her view that although this is Government owned land, and the women have got legal rights to it, their actual possession of the land would only be possible once these structural obstacles in the form of male dominated powerful caste groups, have been faced and overcome. In that sense, the filing of the case, and the citing of the State Government as a party along with SEWA, have made SEWA's position stronger. No longer was the women's claim being treated as only a women's issue, but the State Government was a party to it. Also once SEWA was able to establish the correctness of their position with regard to the land in Metaal, it would have the positive impact of strengthening their position in the other villages.

One is compelled to say that SEWA's hopes have been

belied by the outcome of the court process as well as the continuing insecurity facing the women. Hopefully, this is only a temporary setback. SEWA, with its superior experience, may be proved right in the long run.

In Ganeshpura, Mehsana district, the women are working on *panchayat* land. This is the most successful of all the SEWA projects that we visited. However, the Vanalakshmi Mandali is not yet registered. Most women belong to two caste groups and opposing caste loyalties has so far prevented registration of the cooperative. There is no legal problem that has surfaced so far.

VIKSAT and Legal Difficulties

An example of a genuine legal problem, a problem that is rooted in the way laws are drafted and contradict each other, is the VIKSAT afforestation programme. G. Raju said that originally VIKSAT saw its role primarily as that of a catalyst, helping others to get involved in wasteland development and also providing necessary training to NGOs involved in organising local populations. They did not organise or implement afforestation efforts directly. However, VIKSAT thought it could be useful to have direct field experience. Therefore, in 1985 it started setting up nine tree growers' cooperatives in Sabarkantha, Mehsana and Bhavnagar districts. Of these, only six could be registered. The two forest G.R.s of 1982 and 1984, however, did not lay down any procedures for implementation. This is where the problems began. The VIKSAT faced problems due to two inherent legal contradictions:

1. The Forest Department can only lease land to a registered cooperative society. On the other hand, the Registrar for the Cooperative Societies wanted the applicants to prove their economic viability and to show a land lease before according registration to them. VIKSAT overcame this impasse by getting a provisional certificate from the Forest Department allotting land to the potential cooperatives. It was only after this *via media* was evolved that the first of the six cooperatives was registered in 1986. It may be noted here that SEWA works on government revenue land and did not face this problem at all. Renana

Jhabvala, however, said that they were compelled to start with registration because they could not get the lease without it.

2. The Forest (Conservation) Act of 1980 (Section 2) was amended in 1988 making the operation of the G.R. of 1984 impossible. We were informed by reliable sources that this amendment was on the anvil from 1986 and officials had been advised not to approve of any lease of forest lands. Section 2 begins: 'Notwithstanding anything contained in any other law for the time being in force in a State, no State Government or other authority shall make, except with the prior approval of the Government, any order directing....'

In 1988, the following insertion was made in Section 2:

> (iii) that any forest land or any portion there may be assigned by way of lease or otherwise to any private person or to any authority, corporation, agency or any other organisation not owned, managed or controlled by government.

Consequently, even after getting the registration under the Cooperative Societies Act 1962, the tree growers' cooperatives still could not get the forest land. In 1987, the lease applications from the cooperatives were forwarded by the Gujarat Forest Department to the Ministry of Environment in the Central Government. It was then suggested that instead of leasing forest land to cooperatives, tree *pattas* should be given to individuals. This necessitated a fresh round of applications. But those too were rejected by the Ministry of Environment in 1988. Ultimately, VIKSAT decided to work on private lands.

VIKSAT helped to register the Khutba village tree growers' cooperative and secured a lease of ten hectares of land. It took one year for the land to be handed over to the cooperative so work could begin only in 1989. Similar to what had happened in Metaal and Baldana, the Khutba cooperative's land was claimed by the Rabharis, a cowherd community of Khutba and Valsda villages. As soon as the work of planting saplings began, the Valsda Rabharis came and uprooted the saplings. At an informal meeting in the presence of the *mamlatdar*, the cooperative agreed to set aside about two hectares of land for the cattle of Valsda. But the Rabharis did not honour the agreement and continued to trespass. The Khutba cooperative had to get an injunction from the court against the Rabharis in order to resume planting. One possible reason for intra-village hostilities in Metaal and Baldana, as perceived by Renana Jhabvala, was that SEWA had begun too big—200 acres is a lot of land. Dhagamwar also felt that the hostility reflected gender-prejudice. The example of Khutba disproves both suspicions as ten hectares is a small patch of land and the cooperative was formed by men.

Some General Issues

After looking at the problems faced by the cooperatives of Baldana, Metaal and Khutba, one is constrained to say that while their disputes have been given a legal colour, they are in fact not legal problems at all. These specious legal battles have arisen out of a tussle to control scarce land resources, whether or not the parties to the dispute have a legal right to them. The problem is how to stop such *mala fide* use of law. However, such manipulation of the legal system is not limited to wasteland cooperatives but is a fairly universal practice.

It is evident that forest land cooperatives did not take off as hoped because of a genuine and typical legal problem created by the fact that each law seems to function in a vacuum. Thus, both the Forest Department and the Registrar for Cooperative Societies wanted the other to proceed first! To satisfy two sets of laws with contradictory demands is virtually impossible. A bill, or at least a G.R., which will solve the contradictions between the two State laws and permit speedy registration of tree growers' cooperatives, seems to be urgently called for.

The second and equally insoluble difficulty faced by VIKSAT arose out of the fact that before forest land could be given to the forest cooperatives, registered after such difficulties, the requirements of two levels of government, State and Central, had to be satisfied.

Here too, the Forest (Conservation) Act (Section 2) could perhaps be amended to provide that where a tree growers' cooperative has been registered as per the Forest Department's stipulations, lease of forest land should be given by the State Government and the Central Government should be kept informed of the same. The NGOs involved in afforestation programmes could perhaps campaign for these legal changes.

Conclusion

For our concluding remarks we return once more to the purpose of this study. We had begun with certain specific expectations. The women oriented wasteland development projects are expected to fulfil several distinct, though related, goals:

—create employment,
—generate income,
—raise the status of women,
—give women access to land,
—afforest areas which have been without tree cover, and
—meet subsistence needs.

How far and to what extent did the projects we visited satisfy these expectations?

To begin with, the projects have generated wage employment for the members of the cooperatives at the time of digging pits and planting saplings. In Ganeshpura, income from the land has started coming in so one may justifiably expect these projects to generate income in a few years.

It should be said, however, that we feel that it is somewhat premature to evaluate or assess the wasteland development projects we visited as none of them (with the exception of Ganeshpura which began in 1985) are more than three years old. Of these, the first two years were drought years and the third and last saw unexpectedly heavy rains in many parts of Gujarat, including some of the project areas of SEWA. Consequently, except for Ganeshpura, there is hardly any plantation to be seen. Everywhere we went the women were busy planting for the third time. In Kokta, it was for the first. It follows that the satisfaction of needs for fuel and fodder from these projects is still a distant dream, and cannot yet be measured in concrete terms.

One doubt, however, arises on account of the short duration of the leases. How are these goals to be met after the lease has expired? Will the government give the women continued access to the land even if the lease is not renewed?

In all four projects the women we met clearly pinned high hopes on their projects. When we asked them, the women replied that they expected to get fuel and fodder from the land, and to get vegetables and fruit to sell and eat.

What about the impact on the status of women? Most important of all to the women was the sense of security and continuity that they got from possession of land even though the yield from the plantation was still to come. According to the women, the great gain was that they would no longer have to run around constantly looking for employment. They would no longer be looking for, asking or begging for work. They now had their own work to do. In Baldana, the women declared that they would prefer to work on their own land even if the wage was somewhat higher outside. An employer could dismiss workers and always hire workers for a short period. But this land was their own for 15 years which seemed a long time to them. In their eyes they were better off and their future had a certain stability. We would be inclined to say that if the women do not have to run after employers, that in itself would lead to a rise in their status, both in their own perception and in the eyes of the village. There would no longer be a pool of surplus and waiting women labourers with no bargaining capacity or staying power. Access to land, even before it yields results, seems to have led to a situation where women's status is likely to rise. All this is still to come. At the moment, the impact is not visible.

One question posed to us was about the types of organisation that would work best in achieving afforestation objectives. This is surely a very complex question, not to be answered very easily. It can be attempted from the viewpoint of the legal requirements and also the actual organisational inputs in dealing with the law.

We saw two models of afforestation. One on *panchayat* land where there were no cooperative societies and the second on government revenue land where there were cooperative societies. In both situations, the SEWA registered a big presence. Its full-time workers and field organisers were in constant touch with the project. The cooperative societies are no less dependent on SEWA for information and management than the *panchayat* land afforestation groups.

According to Jhabvala, the advantage of having a village cooperative at the helm is that the whole village feels involved and empowered. This perception is belied by the sequence of events witnessed in Baldana and Metaal. Yet it is also true that in the two villages where the projects were situated on *panchayat* land and with *panchayat* cooperation, the women faced no problems over access to land, though there was a caste-based problem in Ganeshpura. On the other hand, an intermediary NGO is better equipped than village-level organisations to provide technical inputs, do the required paperwork and liaise with the government and officials. It is not without significance that the entire litigation in Metaal was handled by SEWA, though it was not a party to the litigation. The telegrams of stay granted to the Bharwads were sent to SEWA. It was again SEWA office-bearers who signed as witnesses to the *samadhan*, not the cooperative or *mandali* office-bearers. The *mandali* members also could not have coped with the problems of engaging and briefing a lawyer.

It was the SEWA Coordinator who corresponded with the Assistant Collector and Collector and liaised with the Gujarat Agricultural University. Even something small but essential like transport would not have been within the easy reach of the cooperatives.

Similar conclusions can be drawn from the VIKSAT experiences. It is notable that the forest land cooperatives were meant to help marginal farmers and landless persons. Yet the long drawn out struggle involved in registering them imposed an unbearable strain on the people which they could only survive with the backing of a well-established NGO.

One reason for NGOs to hand over any projects to cooperatives and ease themselves out is the question of management and efficiency. Each cooperative is a unit in itself. But if an NGO is compelled to run all the village projects, there is a limit on its capacity to expand its work. The SEWA style has always been to encourage their projects to become independent and separate. It is clear, however, that it is unrealistic under the prevailing conditions to expect afforestation or other projects to be taken up by village-level cooperatives from the very start on their own.

In our discussions with her, Renana Jhabvala agreed with us that the ideal situation would be for a non-governmental organisation, which is already working in the area, to take up afforestation and then gradually hand it over to a village cooperative of tree growers. However, as the existence of a village tree growers' cooperative is a prerequisite for getting lease of government wasteland under the 1987 G.R., SEWA had no option but to start with registration of cooperatives. NGOs involved in afforestation could perhaps get together to decide whether to ask the Government to modify the requirement of prior registration of cooperatives.

The success of women's participation in social forestry programmes depends on yet another factor. As Anil Shah of AKRSP explained, and indeed as has been the general experience, even if a programme is planned with a large degree of women's participation built into it because women are not represented in the decision-making bodies, the programme gets taken over by the men. The women have only a token representation.

Therefore, it would appear that the only programmes that can involve women in a large way both in decision-making processes and its execution are programmes designed exclusively for them; in other words, the SEWA model.

How far will the wasteland development projects achieve the goal of afforestation? After the expiry of the first long lease, the G.R. provides for only one-year lease extensions at a time. This period is neither enough to plan for tree plantation nor for their nurture. It will also induce a feeling of insecurity amongst the members of the cooperatives and even a sense of alienation from the land. The members may well not be willing to put up a fight for a plot that, after all, may be taken away from them in a few months. They may also not be keen to protect the trees or plant slow growing trees.

We are aware that what we say can be perceived as leading to the proposition that the land should be leased permanently to the cooperative. This would be like saying that forests should be handed over to the people. We are also informed that there is a certain amount of controversy about this proposition. We are conscious that if the land were to be given on permanent lease to a cooperative of otherwise landless members, they may be more attracted to horticulture or even agriculture instead of forestry—the women of Ganeshpura were looking forward to the mustard seed crop and those of Kokta and Baldana to their vegetable harvest. A *via media* may be to give long leases of 50 years or more.

Alternatively, the leases should be renewed for 15 years at a time. Rules for *not* renewing leases should also be framed, renewal being the norm. The cooperative should certainly be given much more than five years to prove its capabilities. The Government agencies will naturally monitor the programme, as they already do now.

In the evaluation of any project the perceptions of those whose lives depend upon it are surely of vital importance. The lit-up faces and cheerful smiles of the women were a sure indication that the women themselves were full of hopes for the future and their hope itself is a major gain of these projects. For those who have nothing, even a 15-year relationship with land is a beginning of significance and promise.

REFERENCE

CSE (1985) *The state of India's environment, 1984–85: The second citizen's report*. New Delhi: Centre for Science and Environment.

6 *Caste, Class, Tribe and Gender: Factors Affecting Women's Participation in Wasteland Development*

NEERA BURRA

INTRODUCTION

This paper aims at reviewing the experiences of non-governmental organisations (NGOs) with women's involvement in wasteland development. It is based on visits to a large number of NGOs in the country which are actively engaged in wasteland development work. These visits, however, were necessarily short and the observations are therefore tentative. But apart from these short visits to some 20 NGOs all over the country, about a year was spent in reading the secondary material on wasteland development and social forestry. While there is a vast amount of literature available on social forestry, wasteland development and other wider environmental issues, much of it deals rather skimpily with the issue of women's participation and involvement. This paper attempts to place women's participation in wasteland development in the larger sociological context of the village society by describing and analysing the different approaches adopted by various NGOs. Caste and class, male out-migration, homogeneous villages versus multi-caste villages and the topography of villages are some of the factors which have a strong bearing on wasteland development and women's participation. There are also wide regional variations which have been documented in this paper. These differences mitigate against a common prescription for women's involvement in wasteland development.

It is widely believed that women have a special relationship with land and forests in particular as they are the gatherers of fuel, food and fodder, not only for domestic but also for commercial consumption (Fernandes and Menon, 1987: 16–17; Shiva, 1988; Saxena, 1989: 15–16; Fernandes, Menon and Viegas, 1988: 252; Fortmann and Rocheleau, 1989: 24). Yet, with rapid deforestation and limited access to land, women are being increasingly deprived of major sources of survival such as twigs, leaves, flowers, fruits, honey and medicinal plants and now have to go further afield to find them (Fernandes, Menon and Viegas, 1988: 116, 124).

Wasteland development is seen as one strategy to rectify this growing imbalance. In the argument in favour of women's special relationship with land and forests, a number of assumptions have been made—that poor women's needs are different from those of poor men; that women's knowledge is greater (Molnar and Schreiber 1989: 2); and that while men want to plant commercial species, women's interests centre primarily around fuelwood and fodder. While these assumptions may well be true in respect of specific communities in particular settings, one question that arises is as to whether these assumptions should be regarded as true *a priori*. In this paper, we will test these assumptions with evidence from the field. The paper also looks at the dimension of caste and tribe, studies the contrast in the needs of the poor and those of commerce and industry, and critically surveys the policy leading to privatising common lands. And finally, an important issue dealt with has to do with the circumstances and contexts that promote or discourage women's participation in wasteland development.

Wasteland Development: Some Approaches

In what follows, I have described in some detail the approaches chosen by different voluntary groups in executing wasteland development projects through the participation of women.[1] The projects described below cover the States of Uttar Pradesh, West Bengal

[1] See also the proceedings of the National Workshop on Small Scale Watershed Development (SPWD, 1988).

Orissa, Andhra Pradesh, Karnataka and Rajasthan. Apart from geographical diversity, the voluntary groups also display significant differences in terms of their background and orientation. On the one hand, we find groups like Dasholi Gram Swarajya Mandal and Jakeshwar Shikshan Sanstha which have emerged from a long tradition of *sarvodaya* and they have been engaged in relief and reconstructive activity in the UP hills for many years before they took up wasteland development; on the other hand, a number of groups made up of people with essentially urban origins have sought to take up such projects quite recently in rural areas.

Rajasthan

Three NGOs in Rajasthan share a commonality of approach and work amidst the same target population. Ubeshwar Vikas Mandal (UVM) is located in Udaipur. It was started in 1984 and works in the Girva and Gogunda *tehsils* covering 45 villages. They have planted on more than 497 hectares of wasteland.[2] Seva Mandir conducts its programmes in Udaipur district as well. It was set up in 1952 though its rural activities started in 1969. They work in more than 350 villages in five primarily tribal blocks of Badgaon, Girwa, Kherwara, Jhadol and Kotda. Their wasteland work was taken up more recently in the 1980s and they have undertaken plantation work in approximately 300 hectares of private pooled land.[3] The Jan Shiksha Avam Vikas Sangathan, commonly known as the People's Education and Development Organisation (PEDO) has its sphere of operation in 175 villages of the Bicchiwara block of Dungarpur district. They have developed 2091 hectares of common lands in 30 villages benefiting 3470 families. They have also taken up plantation on 430 hectares of private land which has benefited 1196 families.[4] Most of the villages these agencies cover are populated by Bhil tribals and the social composition of these villages is largely homogeneous. Though

[2] See UVM (1989).

[3] See Khetan (1989).

[4] See PEDO (1989).

we shall return to the implications of homogeneity later in the paper, suffice it to say here that it leads to fewer social strains in the implementation of programmes. The terrain is inhospitable and mostly barren, located as it is in the Aravalli hills of Rajasthan. Women's involvement in wasteland development has been spurred by the demographic fact of significant male out-migration. Lack of employment opportunities forces tribal males to travel mainly to Gujarat in search of a livelihood. It needs to be said, however, that there are varying emphases upon the extent to which women were explicitly or exclusively targeted amongst the groups visited. During our discussions with women's groups organised by Seva Mandir, it became apparent that the general approach of the agency to make women conscious of their rights had truly engendered confidence amongst them. The agency believed that only when such awareness dawned could development programmes succeed.

Though UVM initially took up wasteland development on private lands that were pooled together, today its focus is upon wasteland development on common lands, whether these belong to the *panchayat*, the Revenue or Forest Departments. Seva Mandir works mainly on private lands but also on common lands. PEDO helps develop common lands but also develops private lands in order to relieve pressure upon common pasture lands. Amongst the Bhils in this area, the distribution of land is relatively egalitarian and almost every family owns an acre or two of degraded land. Moreover, there are in addition vast tracts of wasteland belonging to the *panchayats* or government departments. What is noticeable in the endeavours of all these groups is the effort made to evolve a consensus in the community as to usufruct rights on the land once it is developed. Negotiations and discussions take place over long periods of time and when agreements are reached, these are frequently recorded in writing to the knowledge of local revenue and forest authorities. Careful mapping of the plots is resorted to in order to forestall the possibilities of disputes arising at a later time.

The following is an illustration of the kind of consensual decision reached. In Upalgaon village, PEDO held a three-day *shibir* to decide how the produce from the wasteland was going to be distributed. An unanimous decision was taken that green fodder would be provided for two animals per household but that poorer households could cut more dry fodder. This kind of community decision-making is often possible if the *falah* is relatively small. In larger *falahs*, there are village-level representatives—50 per cent men and 50 per cent women.

Uttar Pradesh

In the Uttarakhand[5] region of Uttar Pradesh, there are three types of forest lands: Forest Department land, reserved forest land (also known as civil/*soyam* land) and private land. During the colonial period when the government took over the forests, people protested because hitherto they had exclusive rights.[6] As a special concession, forests near villages were demarcated, separate bodies known as *van panchayats* with their own committees were set up and the villagers were given exclusive rights over fuel and fodder from these *van panchayat* forests. But over the years, due to population pressures and open access without proper management, the *van panchayat* forests have become almost completely degraded and the people are now cutting fuel and fodder from reserved forests.

In Gopeshwar, the district headquarters of Chamoli district, I visited the Jakeshwar Shikshan Sanstha (JSS) and the Dasholi Gram Swarajya Mandal (DGSM). Anand Singh Bisht, the founder of JSS, had been part of DGSM which has become famous for the Chipko movement. (Chipko means to hug and the movement is known as the Chipko movement because the women in the area hugged the trees and refused to let the contractors cut them). The ideology and approach of both these organisations are exactly the same. JSS is a splinter group of DGSM. Chandi Prasad Bhatt is the founder of DGSM which was formed in 1964. JSS was founded in 1978.

The Chipko movement against the felling of trees has been well documented (Bhatt, 1990: 6–11; Gadgil, 1989: 33; SPWD, 1984: 4–9;

[5] The eight hill districts of Pauri, Chamoli, Tehri, Uttarkashi, Dehradun, Almora, Pithoragarh and Nainital comprise the Uttarakhand region.

[6] Ramachandra Guha (1989) gives an excellent historical account of the traditional rights the people of Uttarakhand enjoyed and how they have protested against state interventions.

Dankelman et al., 1989: 11–17; Freerk, 1986: 22–23).[7] An interesting fact is that both these organisations took up plantation and protection work simultaneously. Anand Singh Bisht and Chandi Prasad Bhatt belong to the *sarvodaya* movement and they have been closely involved with development activities in the rural areas. In 1970, there was a devastating flood on the Alaknanda river washing away a large number of people and their homes. Before government relief work could be mobilised, Bhatt and Bisht organised groups of relief workers. After the flood had subsided, they did a survey using local people to identify the causes of the landslide which led to the flood. Their findings were that where there had been large-scale felling of trees, there was much greater damage done than in areas where there was better tree cover (Guha, 1989: 156). This information spurred them on. They also realised that women now had to travel miles for collecting fuel and fodder and therefore, organising women was absolutely necessary.

Most of the plantation work that I saw was around the village. When I asked some of the women what would happen if the Forest Department did not allow them to cut the trees later or got the trees cut for timber, the response everywhere was unanimous: 'Let them just try!' I found this in all the three villages of Koteshwar, Bacher and Papadian that I visited.

Bacher and Papadian villages come under the overall supervision of DGSM which works in 70 villages. Chandi Prasad Bhatt, Chakradhar Dutt Tiwari and Kalavati Devi took me to Bacher which is 35 kilometres from Gopeshwar and in the interior. I had long discussions with Kalavati Devi, the President of Bacher's *mahila mangal dal*. She said that in their village they have formed the first *mahila van panchayat*. Earlier *van panchayats* were completely controlled by men. The *sarpanch* and the seven *panchas* were all men and women rarely attended the meetings. But after forming the *mahila mangal dal* in Bacher, the women insisted that *van panchayat* meetings could not take place unless all the women were present. Recently, some women discovered that the *sarpanch* of the *van panchayat* was involved in the illegal felling of trees from the village forest. They blocked the road from the village and called an emergency meeting. Since women

[7] For a detailed social history of the Chipko movement, see Guha, 1989: 152–83.

outnumber men in this village, they were able to elect a woman as the *sarpanch* . In Bacher, four *panchas* are women and four are men. I was told that when the next elections take place, more *van panchayats* will be taken over by women. Elections take place every five years according to the *Van Panchayat Act*.

One of the factors that has motivated women into getting so involved is the manner in which the organising work is done. Both the JSS and DGSM regularly organise seminars in different villages. This is done particularly when some work such as starting a plantation is about to be taken up. These camps are organised three or four times a year and each camp has 250–300 participants. Seventy-five per cent of the participants are village women. Both the JSS and DGSM also invite outside journalists, students, environmentalists, social workers and local government officials. During these seminars, village development is discussed and a master plan is decided upon. All the participants help with *shramdaan* to dig pits and construct boundary walls. Both the DGSM and JSS find that this *shramdaan* is extremely useful as it hastens the process of completion of work and also gives the village some visibility. Local journalists have documented these camps and give the village women a lot of publicity in the local district papers. Whenever there is anything published about the Chipko movement, it is shown to the women and discussed in the villages. This again builds their self-esteem and confidence. When DGSM was awarded the Environment Award by the former Prime Minister, Rajiv Gandhi, it was received by about 30 Presidents of *mahila mangal dals* who came to Delhi on this occasion. This has boosted their morale as well as made others want to improve their plantation and take up protection work around their villages. In vast tracts on the Garhwal hillsides, by merely controlling access, plant life has regenerated to such an extent that the amount of plantation work necessary is also reduced. I was told that in the Garhwal hills there are many such organisations that are working with the people and are having a visible impact upon the environment.

As examples of work at grass-roots and people's participation, the performance of JSS and DGSM is outstanding and they provide excellent models for replication. Some of the main features of these NGOs are that the leaders belong to the area and can and do identify themselves with the people. A substantial amount of time has been spent by these organisations in helping rural women organise. They have maintained a very low profile and have only intervened during a crisis. They have deliberately decentralised and allowed initiative at the grass-roots to grow. They foster a genuine belief that women need to be mobilised so that they can assert themselves; once a group has developed, their involvement becomes that of observers. Their personal style of living is Gandhian and austere.[8]

I also had occasion to see the work of an NGO in Almora district which is contiguous to Chamoli district. I found very little enthusiasm for plantation work among the NGOs in Almora even though people admitted that the socio-economic position of women in Almora and Chamoli was similar. Their problems with the Forest Department are similar, their needs for fuel and fodder are the same and the hardships due to non-availability of these essential commodities are common. My own understanding in the few days that I spent in Chamoli was that it was a question of difference in approach. The Chamoli approach was essentially one of building up women's consciousness about issues that affected their daily lives; if the women wanted toilets or creches or to lead an agitation against the liquor trade, then that is what they were helped to do. The idea of protecting their forests and planting came up only when their collective identity was established.

West Bengal

The Centre for Women's Development Studies (CWDS) has been working with tribal women, mainly Santhal, in the Bankura district of West Bengal since 1980. Wasteland development was one of several income-generating activities they started in the area. However, wasteland development as a major activity for improving the income levels of women was taken up only in 1987 when the men of these families agreed to donate their private wasteland to the *samitis*. The women have planted mainly *arjun* and *asan* plantations for rearing *tassar* silkworm cocoons—*tassar* silk worms feed on the leaves of *arjun* and *asan* trees and the thread is woven into silk.

[8] Chandi Prasad Bhatt was awarded the Ramon Magsaysay Award for his outstanding work. Like an ordinary villager he either walks or travels by bus and lives in one room at the office.

The CWDS project, executed by ILO,[9] was one of the earliest wasteland development projects which saw wasteland development as a potentially viable economic activity which could generate significant and sustained employment opportunities.[10] With sheer hard work, the women have very successfully raised plantations on more than 300 acres of private pooled land which has been donated to their organisations. These plantations have generated about 150 days of employment per annum for more than 600 members of 15 *samitis*. Since this work is in addition to what the women can manage by working on their own or others' agricultural lands, it has meant that they no longer have to go on *namaal*.

One of the first activities CWDS started was the mobilisation and organisation of the local women. This remains one of the greatest strengths of the project. Each village has a women's *samiti* and each *samiti* has representatives in the apex body called the Nari Bikash Sangh (NBS). NBS is now the main catalyst and mobiliser and most activities are carried out under their aegis. CWDS is a facilitator supporting NBS from the outside and providing technical and other assistance as and when necessary. When requests are made by women in other villages to start wasteland development work in their area, it is the NBS leaders who go to the village, assess the area of land available and motivate and organise the women.

CWDS and NBS have become very cautious in taking up new areas even though the demand for assistance in wasteland development is growing. Earlier, when CWDS was first experimenting with wasteland development activities, they would organise a large number of women, regardless of the area of land available. Sometimes, multi-village *samitis* were formed. This later created problems because women living in the vicinity of the plantations were benefited more than others. They also found that the volume of employment generated by wasteland development could not absorb as many women as were keen to find work. As a result, there was some disillusionment and the plantations did not turn out to be viable as planned. When CWDS decided to concentrate

[9] See International Labour Organisation (1988) *The Bankura Story: Rural Women Organise for Change*. New Delhi: International Labour Organisation.

[10] The total membership of the *samitis* is 1,663. There are a total of 19 registered *samitis*. The earlier phase of the project dealt with employment opportunities and did not focus upon wasteland development. The figures in the paper regarding membership, number of *samitis* and employment generated relate to the wasteland project only up to May 1993.

upon wasteland development as their main focus in 1987, they changed their norms. They now promote smaller *samitis* with fewer members so that the ratio of women to land can ensure sufficient employment for each member. Now they are also trying to negotiate for use rights to vested (ceiling surplus) land for the *samitis*.

An interesting new development has been the increasing involvement of the *samiti* women in the Forest Protection Committees (FPCs) in the area. The West Bengal Forest Department has since 1982 been organising FPCs in three districts of West Bengal—Bankura, Purulia and Midnapore.[11] In the Ranibandh range where most of the CWDS's project villages are situated, there are 40 FPCs protecting a total of 8,777 hectares of degraded *sal* forest. Under this scheme, if the village volunteers to set up a FPC and protect its forest for five years, the village is entitled to 25 per cent of the value of the timber at the time of the thinning of the forest and another 25 per cent of the timber at the final felling. In the meanwhile, villagers can collect as much non-timber forest product as they want without any interference from the Forest Department.

We visited Makhno village which is 14 kilometres from Jhilimili, the base town for the CWDS project. Makhno is where one of the first FPCs was set up and where the villagers are protecting 153.58 hectares of *sal* forest. In this village, the members of the FPC are all men though they admitted that the real protection was done by the women. It is the women who go to the forest to collect twigs, *mahua* flowers, fruits, *sal* seeds and leaves and graze the animals. Because of their regular presence in the forests, they are the ones most likely to be aware of outsiders encroaching on forest resources. At the meeting we had in the village, the women were nowhere to be seen.

But in some CWDS *samitis*—like the one in Kodapada village—the members have set up FPCs in which 50 per cent of the members are women. More recently, NBS held a meeting of all its members and passed a resolution that all FPCs should have at least 50 per cent women members as it was women who were *de facto* protectors of the forest. They received a very positive response from the Forest Department which then issued a notification to all Range

[11] For details about the FPCs, see Malhotra and Poffenberger, 1989; *Wastelands News*, 1990: 27–28.

Officers in the area that in future all FPCs should have at least 30 per cent women members. The *samiti* members, under the overall guidance of the NBS, plan to form FPCs and also help to prepare micro-plans for their areas. All this is of course possible because of a spearhead group like the NBS which is made up of local women who are experienced in leadership and in playing a catalyst's role. While CWDS has no immediate plans to withdraw from the area, their role now is mainly to liaise with the outside world. Most major decisions, however, are taken by NBS leaders.

Orissa

The People's Rural Education Movement (PREM) was started in 1980 in the Mohana block of Ganjam district of Orissa. They work in 12 blocks in Ganjam district amongst the Kond and Saura tribal groups. They have developed 1370 hectares of wasteland in the four blocks of Nuagada, Mohana, Surada and Rangelunda. This agency is attempting to conscientise the tribals about their rights and is trying to ensure that they get minimum wages. They are also involved with a number of other activities in the educational and health fields. Slash-and-burn (*bogodu*) cultivation is the predominant agricultural practice in this area. PREM became involved with wasteland development as a result of the problems faced by tribals engaged in *bogodu* cultivation. Traditional agricultural practice consists of slashing and burning the vegetation in March, mixing the seeds of *channa, ragi* and a variety of pulses, sowing them in the months of April, May and June and harvesting them in September. Wage employment for the rest of the year is uncertain, erratic and does not offer a decent livelihood. By cutting bamboo illegally from the forest and selling some non-timber forest produce they can add to their meagre income.

The Forest Department fines every family Rs. 1100 if they are caught slashing and burning forests. We were told that every family in the region is caught at least once every two or three years and members have to spend some time in jail if they do not or cannot pay the money. In one village we visited, a man had sold four goats to pay the fine and had begun the practice of rearing goats in order to deal with such contingencies. Entire villages are now trying to build the corpus of a village fund to pay these fines. PREM actively assists them in this endeavour and hires advocates to represent the tribals in court.

PREM came to the hard-headed economic realisation that *bogodu* cultivation per acre yields only Rs. 8000 if the harvest is good, whereas by planting a mix of pineapples, mango, cashew, banana and papaya, an acre could yield about Rs. 18,000–20,000. PREM promotes such plantations in order to increase the income-earning opportunities of the tribals and they have deliberately stayed away from concentrating on fodder and fuel species for these are not related to the felt needs of the tribals. By encouraging such mixed fruit plantations, PREM hopes that the consequent economic gain will discourage slash-and-burn cultivation. Revenue and Forest Department officials see the growing popularity of fruit plantations as a check upon *bogodu* cultivation and PREM is promoting village-level groups to enter into agreements with the concerned governmental authorities. Out of 12 block-level organisations sponsored by PREM, eight have taken up plantation work in 72 villages. About 800 acres of fruit trees have matured out of the 2,500 acres where PREM has undertaken plantation. Ironically, the sale of produce from wasteland development is being used to pay the fine levied for cutting the forest.

Karnataka

One of the most interesting women and wasteland development projects I saw was in Dharwar district of Karnataka State. It is run by the India Development Services (IDS). IDS was started in 1979 and has two project areas, one in Dharwar *taluka* and the other in Ranibennur *taluka* of Dharwar district. They are working in 70 villages in Dharwar *taluka* and 21 villages in Ranibennur *taluka*. Dharwar *taluka* is resource-rich and has a good rainfall while Ranibennur is drought-prone.

The women and wasteland development work in Dharwar *taluka* is quite different from that in Ranibennur. In Dharwar, since there are still vast tracts of forests and forest lands, the approach of the IDS has been to link the local people with the

Forest Department so that they can get the right to collect non-timber forest produce. IDS has a very strong field team in the village which holds regular meetings with the women. The level of decentralisation was impressive: IDS staff live in the village so that they can deal with day-to-day issues. It is one of the conditions of joining the IDS that the staff at every level, including the Project Coordinator, live in the village.

I visited the IDS wasteland project in Medleri village in Ranibennur *taluka* as well. In this programme, they run *kisan* nurseries. In five years (1984–1989), they had sold 14,000,000 saplings to the Forest Department. The Forest Department had provided the seeds and the bags and had later purchased the saplings. They paid Rs. 250 per thousand saplings. About 400 women participated in 12 villages. Each village has nursery groups or *sanghas* of 10–15 members. Nursery-raising is done in groups.

In Ranibennur, IDS has been involved in a variety of ways with women and wasteland development. They have a strong presence in the area as their field office is located in Medleri village near Ranibennur. Since there is an acute shortage of fodder and fuel, most of the efforts of the IDS have been directed towards these issues. Their wasteland development work started after they had established a presence in the area and organised women's groups over other concerns. They have also organised women's dairy cooperatives.

In Aklaspur village, there are 1400 acres of land under the control of the Forest Department. About 1000 acres are under the plantation of *anjun* which is a drought-resistant fodder species. The plantation was not being looked after by the Forest Department and the villagers were indiscriminately cutting the leaves and branches. The women's dairy cooperative society wrote to the District Forest Officer about the rapid degradation of the plantation. Since the people were cutting off branches as well as the tender leaves, no *anjun* seeds were available for nursery-raising in the area. The Forest Department responded immediately and held training programmes in which about 20 people participated and were shown how the leaves should be cut so that the trees could grow as well. The women's dairy cooperative society has been given 100 acres to manage and the regeneration of this tract was remarkable, with mere protection. In 1989 the villagers were able to collect *anjun* seeds which they sold to the Forest Department. Because of the initiative taken by the women's group, they were given responsibility for the management of a large plantation. The idea is that once this tract regenerates, it will be opened for access and another portion of the plantation will be given to them to protect. In this manner, the 1000-acre *anjun* plantation will be regenerated. During recent drought years, this *anjun* plantation saved the lives of hundreds of animals.

What is remarkable about the IDS is the variety of interventions they have made. These range from the training of women with the help of the Forest Department in how to use trees without damaging them to raising nurseries to planting and to becoming aware of the larger contexts in which their struggles are placed. Later in this paper, we will describe major conflicts that IDS became involved in as a consequence of their opposition to commercial and industrial interests and in defence of the needs of ordinary people.

Andhra Pradesh

The Bhagavatula Charitable Trust (BCT) was set up in 1977 and was one of the first organisations in the country to take up plantation work on wasteland. Their objective was to show that if properly nurtured, nothing was 'wasteland', only 'wasted' land. The first experiment was with 50 acres of revenue wasteland which was given by the Government of Andhra Pradesh to BCT on a 20-year lease. The Panchadarla Farm, as this is known, has been widely cited as a very successful experiment in wasteland development. They have other farms developed on wasteland, notably the Gokiwada farm (on 44 acres) which belongs to 99 farmers and the Gorrela Dharmavaram farm which is on 58 acres and belongs to 110 poor families. When the National Wasteland Development Board (NWDB) was first set up, Dr. Parameswara Rao, the founder of BCT, was made a member. BCT works in Yellamanchilli *taluka* of Visakhapatnam district.

I went to visit the Panchadarla Farm. The farm is well-planned and well-maintained. There was a Farm Supervisor and a Manager. The plants were well looked after. The trees were laden with fruit. It was explained how intercropping had taken place. The fodder was used to feed the cattle in the cattle-breeding centre. All the biomass was fed into a plant which generated biogas. In fact, nothing was wasted on the farm. The farm had been planned in such a way as to obtain the maximum yield from the land. BCT had gone to the extent of demonstrating different ways of planting the same species, e.g., the same species of jackfruit were planted in different ways in the same area on the same soil on the same day. One was planted the way in which the villagers normally plant it by simply digging a pit. Another sapling was planted but a large round trench was dug around the tree to harvest the water and all the dried leaves from the tree were put into the trench. One was bearing fruit and the other one was growing with very few leaves and no fruit. I was told that this is the way in which the villagers who visit are taught improved techniques of planting. In addition, BCT has taken a lot of trouble to get better varieties of cashew from Kerala, papaya plants from Malaysia and silk cotton from Australia.

This was also the case with the Haripuram farm where the trees were laden with fruit and there were also minor timber trees. Fodder was grown for the animals on the farm. The produce was more than sufficient for all the workers of BCT and their families and, in addition, surpluses were sold. BCT also runs the Gorrela Dharmavaram farm, which they call an all-women's farm, where they have registered a company. Only women can be shareholders and only women can be directors of the company. This is a farm of 58 acres of wasteland pooled together by 110 families of Gorrela Dharmavaram village. I was told that the men had given the land to the women. When I visited the Gorrela Dharmavaram farm, I found that the land was very degraded. In some places there were 20-feet wide gullies and it looked as if the earth had been split open by an earthquake. The plantation work was taken up in 1984. The trees did not seem so big from a distance but like the Panchadarla and Haripuram farms, the place looked well-planned. There were men ploughing with tractors and the work of pit-digging and planting was going on, supervised by one man. I spent some time talking to the women who were working on the plantation as well as to the Supervisor. Later I had the opportunity of speaking with BCT's Chief Agricultural Officer but I did not get the impression that women were involved at every stage of planning and implementation of the project. The women were very active in the nursery-raising work. But the plantations were developed by BCT and then handed over to the *mahila mandals*. There seemed to be an underlying assumption that to make an enterprise economically viable, it had to be technically and scientifically sound and that was only possible if experts were brought into the picture at an initial stage. Once the teething problems were over, then the women could run it quite efficiently. The distinguishing feature of BCT's involvement in wasteland development is the technological and managerial input. The sheer professionalism of the effort was evidenced by its clearly visible success on the farms one saw. Reportedly, the farms are run so well and so profitably that they subsidise other activities of BCT. Careful planning of farm activities has resulted in a highly efficient use of land.

Social Homogeneity and Conflict: Some General Observations

On the basis of the visits I made to the field, as also upon a reading of the literature, one conclusion that struck me forcefully was that those projects appeared to have better chances of success where the village was homogeneous with most inhabitants belonging to a single caste or tribe. Conversely, in a multi-caste village or where there are heterogeneous social groups, wasteland projects are likely to be less successful (Freerk, 1986: 26–27; Chambers, Saxena and Shah, 1989).

Tribal communities, usually located in hilly areas, display a reasonably equitable pattern of distribution of land. In many projects visited, almost all the families owned a few acres of land even if the land was degraded. Seva Mandir, UVM and PEDO are all agencies that work amongst the Bhils in the Aravalli hills of Udaipur and Dungarpur districts of Rajasthan. A common economic and social background and the absence of conflict between different groups are factors that predispose projects towards success in securing the participation of the people. Similarly, most of the villages taken up under the CWDS wasteland development project in the Bankura district of West Bengal are populated primarily by Santhal and Bhumij tribals. Chambers, Saxena and Shah,

(1989: 156) point out that one of the factors in Sukhomajri's success was that it was a homogeneous village, practically of one caste (Gujjars). Only two out of the 59 families are Jats.[12]

In Orissa, both Gram Vikas and PREM work only with tribal groups. In Tamanna village, where Gram Vikas has helped set up a plantation under its social forestry programme, it works only with the Kerandimal Kond tribals, although there are at least 20 non-tribal households belonging to a Scheduled Caste community. The Scheduled Castes are not part of the social forestry project because the government scheme is meant only for tribals. Some of the funds that these NGOs receive come from the Tribal Development Department and can therefore only be used for helping tribal groups.

However, Anthia Madiath of Gram Vikas gave us another reason as to why they had not included Scheduled Castes in their wasteland development project even though economically speaking, the Scheduled Castes were just as poor. She explained that in this region the Scheduled Castes have traditionally been used by the caste Hindus to exploit the tribals and the enmity between the Scheduled Castes and Scheduled Tribes is so great that it is not possible to integrate them into the same programme.[13] Apparently, this is another reason why PREM only works with tribals.

In the ILO project run by SEWA in Ahmedabad district of Gujarat, the target group of women belongs largely to the Scheduled Castes.[14] They are Harijans and do not own land at all. In the three villages (Dumali, Baldana and Metaal) where the project is under way, conflict with the Bharwads, a community of shepherds, has created serious obstacles in the progress of the project. The Bharwads resent any attempt to improve the status and lot of the Harijan women. In one village, the conflict could be resolved only by surrendering part of the land to the Bharwads. The question is now being debated by project authorities as to whether beneficiaries should be selected by reference to their land status rather than caste though there is a close correspondence between not owning land and belonging to the lower castes.

[12] See also SPWD, 1984: 2.

[13] Fernandes and Menon's study (1987: 48) in the same area of Orissa shows that the Scheduled Castes are somewhat worse off and earn much less from most activities than the Scheduled Tribes.

[14] For details about the project, see Dhagamwar and Thukral, 1990.

In another SEWA-run project in Mehsana district, conflict has erupted between women who are Harijans and those who are Prajapatis, a higher caste. The project staff rejected a demand to divide the plantation into two plots so that each caste group could work on its own sub-plot. In the face of this rejection, the women have devised an informal system of working in relays. The day the Harijan women come to work, the Prajapati women do not and vice Versa.

In contrast to the instances cited above, the work being done by JSS and DGSM in Chamoli district of Uttar Pradesh is successful despite the fact that many of the hill villages in the area have a mix of castes with Brahmins, Thakurs and other caste Hindus along with Harijans. Both these organisations are of the *sarvodaya* persuasion and have been involved with many other development activities for a number of years. They have made a very conscious effort to break caste barriers and I was told that notions of purity and pollution are not so strictly observed in villages where the women have been organised. When Harijan and upper caste women go outside their villages to attend meetings, camps and workshops, they mingle relatively freely, even sitting and eating together. However, when they return to the village, a certain distance would have to be maintained in deference to the larger social system. The Harijan settlement was separate from the main village and if a meeting of the *mahila mangal dal* took place in the house of an upper caste woman, the Harijans would sit apart. But as far as the need for protecting and planting the forest was concerned, there was a unity of views which overrode the divisiveness of caste.

Topography and its Impact

Another factor which makes a great difference to the success of a wasteland development project is topography. As Chambers, Saxena and Shah (1989: 161) point out:

> The topography of the upland villages makes their common lands visible from most of the dwellings, so that any unauthorised felling cannot escape notice. In contrast, the area of a flat

village in Central-South India may well be spread over 5–10 kms in one direction which makes it easy for 'free riders' to escape undetected.

In Ghati village in Udaipur district, the Patels and the Adivasis decided to take up afforestation work when they found that stone and rubble were coming down the hillsides onto their private lands because of deforestation. Although they have hired a watchman, it is easy for people to keep an eye from their fields below. Another advantage that hill villages have is that they are so remote that when an outsider visits, everybody knows about it. It is only when the felling of trees is done with the connivance of local people that a problem arises. The President of the Bacher *mahila mangal dal* in Gopeshwar told me that for a long time their *van panchayat* trees were being cut with the connivance of their own villagers and the *sarpanch* of the *van panchayat* was fully involved. When the village women discovered this, they blocked the road and refused to let the truck pass; since there was only one road out of the village, it was possible to prevent a vehicle from entering the area. In Medleri in Dharwar district, the women tried to stop the trucks from leaving the village common land with the eucalyptus poles but were totally helpless as they did not know when and from where the trucks would come. Being a multi-caste village in the plains with acres of common land, it was difficult to monitor what was happening and there were too many different people to contend with.

It is also true that when people are living in remote areas, they are forced to depend more upon each other, thus to some extent neutralising divisive influences.

Gender and Plantation: Are Women's Needs Necessarily Different from Men's?

A widely held belief is that women's needs from land are different from those of men. It is claimed that while men want to grow commercial species like eucalyptus or teak, women prefer the broad-leaved species which meet their subsistence needs for fuel and fodder (Shiva, Sharatchandra and Bandopadhyay, 1981; Fortmann and Rocheleau, 1989: 24; Chowdhry, 1983). N.C. Saxena (1987: 154) also agrees that women's choices are different from men's. He illustrates his case with examples from the Chipko movement in which women wanted to grow broad-leaved species for fodder whereas men wanted commercial species and fruit trees which would bring in cash incomes. In fact, many experts (Dankelman et al., 1989: 13–18) who have used this line of argument have cited the Chipko movement as the main example of women's collective strength and desire to protect their environment. The instance most often cited is the hugging of trees by the women of village Reni in Garhwal. On that particular day, there were no men in the village as they had been called to the district headquarters on the pretext that they would get their compensation money that was long overdue to them from the Government. When the men were away, the contractors conspired to go to the village forest and fell the trees but when the women realised what was going to happen, they hugged the trees and refused to let them be cut. The contractors had to go away finally. There are repeated references to this incident in the literature. Much has been made of this episode and the supposedly primordial connections of women with nature glorified. The Chipko movement has symbolised for many the first grass-roots expression of a feminist ideology.

But Ramachandra Guha's seminal work on the Uttarakhand region looks at the movement in its historical and cultural context. Guha (1989: 175) says that the Chipko movement cannot be romanticised as a '... romantic union of humans, especially women, with nature' (ibid.: 173). He suggests that the movement is not a feminist movement as it is sometimes made out to be.[15] The stereotypes are grossly inadequate '... interpreting the origins, idiom and trajectory of the movement' (ibid.: 173–74). According to him, 'Chipko is only one, ... in a series of protest movements against commercial forestry dating from the earliest days of state intervention' (ibid.: 174). Guha (ibid.: 175) reiterates:

> ... despite the important role played by women, it would be simplistic to characterise Chipko as a feminist movement. In several instances, especially the early mobilisations at Mandal and at Phata, it was men who took the initiative in protecting forests. Women came to the fore in Reni, when in the contrived

[15] Skutsch (1989: 64) takes a similar view.

absence of menfolk they unexpectedly came forward to thwart forest felling. In other agitations, such as Badyargarh, men, women and children have all participated equally. Dungri–Paintoli is the only instance of an overt conflict between men and women over the management and control of forest resources. As such, even at the level of participation Chipko can hardly be said to constitute a women's movement.

There are other studies, notably one undertaken by Gunilla Olsson (1988) in Orissa, which do not support the assumption that women's needs are always different from those of men. In her evaluation of the Orissa social forestry project, she came to the conclusion that there are basically no differences in the preferences of men and women with respect to the species to be planted. Her experience in Orissa was that 'trees are primarily grown for their cash income value .... The priority, expressed by both women and men, for which the money thus gained was to be used, was the creation of an asset—buying paddy land' (ibid.: ix). Moreover, there were also no real differences in the way in which men and women used produce and utilised profits (ibid.: 36).

My own field experience was similar to that of Gunilla Olsson. The women I spoke to did not see sharp differences between their needs and those of their men. In a meeting organised by Seva Mandir in Dechra village, we asked the women which species they preferred and some women said they wanted to grow eucalyptus because it had a good market. But the NGO representative told us that they had discouraged the planting of eucalyptus and the women had accepted their advice.

Some of the popular species in the area were bamboo, *dhabdra* and *khedwa* and both men and women were keen on planting these. Bamboo poles were good housing material, *dhabdra* wood was excellent for making ploughs and *khedwa* produced a resin which is a necessary ingredient for making Indian sweetmeats. Most of the women (who were Bhil tribals) said that there was no difference between their needs and those of their men. According to them, men also have subsistence needs for building and repairing huts, making ploughs and so on.

In Garhwal, the home of the Chipko movement, women are growing fruit trees in a big way and both DGSM and JSS encourage women to grow them. They have also tried to help them with marketing the fruit through the Garhwal Nigam (a public sector agency) but with little success so far.

It has become part of the conventional wisdom to assert a link between the role of women as nurturers and protectors and their superior ability to foster and maintain saplings, plants and trees. This alleged connection is seen to stem from the needs of women to provide fuel for the family and fodder for the animals. It is therefore argued that women are more concerned about the environment and its protection and hence the rationale for designing wasteland projects in which women are the main target group. The empirical question that needs to be raised is whether there is any evidence from the field in support of the view that women are better than men at fostering and maintaining plantations.

In conversation with the staff of Seva Mandir, it appeared as if their experience has been that as far as nursery-raising and plantation work are concerned, there is no significant difference in the quality of work done by men and women. In some places, men were just as committed to plantation work and in some areas, even more committed than women. They found that in socially backward areas—such as Badgaon block of Udaipur district—women were not very responsive when offered work of this nature. Most of Seva Mandir's *van pals* are men. Recently, some women have come forward and now there are five or six women. Both male and female staff members of Seva Mandir were of the firm belief that, as far as the nurturing and protection of wasteland was concerned, there was not enough evidence to show that women are better than men or that their needs are different.

One of Gunilla Olsson's observations about the Orissa social forestry project was that women's involvement in nursery-raising was marginal. According to her, 'It is a popular myth that the majority of nursery workers are women and they are preferred because they are particularly careful and meticulous in handling seedlings. The employment data of the Orissa social forestry project shows clearly that the main workers were men' (Olsson 1988: 19).

An experience I had at Gram Vikas in village Gaon in Ganjam district of Orissa is worth recording. In a plantation plot, one portion was exclusively assigned to women and special grants had been received by the agency for the work to be done for that portion. In the rest of the plot, plantation was done by a mix of men and women. There was no perceptible difference either in the choice of species that were planted or in the quality of the plantation. At best, one could say that men would work only for a wage whereas women might, in the absence of alternative employment opportunities, maintain their plantation even without a wage. When I visited, some women were marking the trees though the project period was over and no more money was available to be paid as wages. But even this is not always the case.

In the plantation at Haldikoti village in Dharwar district of Karnataka, I came across an instance of men, women and children all working voluntarily and without a wage in exchange for usufruct rights over the *muttoga* plants they were growing. The leaves of the *muttoga* trees were used for making leaf plates which could then be sold.

This writer feels that though there is a case for organising wasteland development projects with women as the target group, that case does not have to rest upon the reportedly inherent ability of women to foster and maintain plantation.[16] The implementation of development programmes in general certainly needs a sharper focus upon women to enhance their income-earning opportunities and status—both of which are considerably lower in most of India when compared to those of men. If there is not so much difference between the needs of poor women and men, the underlying conflict, it is suggested, is really between the needs of the poor on the one hand, and those of commerce and industry on the other.

The Poor versus Commerce and Industry

In Chamoli district of Garhwal, when the local people wanted permission to cut some ash trees for making an agricultural implement necessary for making ploughs, they were denied permission. But at the same time, a sports goods firm in Bareilly was given a contract to cut the ash trees for making tennis rackets (Agarwal and Narain, 1990: 38; Shepard, 1988: 28–29; Agarwal, 1988: 18). In different parts of India, it has been well documented as to how the

[16] Skutsch (1989: 64–65) discusses this issue at length as to whether there is enough evidence to show that women's needs are different from those of men.

interests of tribal communities have been adversely affected by commercial exploitation of the forests and the nefarious role of contractors (Mistry and Kharadi, 1990; Mistry, 1987; Jha, 1988: 17; Singh, 1989: 63; Shah, 1990: 2–14).

A fairly recent and well-documented conflict of interest has been reported from Karnataka where there has been a long drawn-out confrontation between the people and the State over rights to common properties. In 1984, the Government of Karnataka formed a joint sector company with the sole purpose of growing eucalyptus for a private sector rayon company on 75,000 acres of degraded land. The people did not want this monoculture of eucalyptus on common grazing lands. There were several protest movements when the people, mainly women, uprooted the eucalyptus saplings and planted fruit tree saplings (Sharma, 1987; Bhattacharjea, 1987a, 1987b, 1988a, 1988b).

Again, in Medleri village in Dharwar district, another common property issue has been raised. In 1981, the *panchayat* took up eight hectares of common grazing land for eucalyptus plantation under the social forestry project. When the people protested, they were told that when the trees would be auctioned, the people would have the right to lops and tops and could also give their pole requirements which would be met. At the beginning of the year, it was decided to auction the plantation to the highest bidder. The plantation was sold for Rs. 50,000 and was bought on behalf of the private company. When this happened, IDS and another organisation, Samaj Parivartana Samudaya (SPS), filed a case in the High Court and obtained a stay order. Probably for political reasons, the *panchayat* disobeyed the court order and authorised felling. The women organised by IDS hugged the trees (as was done in the Chipko movement) to prevent them from being cut, but felling continued. The local police and the bureaucracy refused to intervene and enforce the stay order. The whole plantation has gone to the contractor. Due to an acute shortage of fuelwood in the area, even the lops and tops were taken away by the contractor to be sold as fuelwood. For the entire week the people watched truckloads of eucalyptus leaving the plantation. They were told that only 2000 trees had been auctioned though, in fact, the entire plantation was cut. The timber value of that was over Rs. 8,000,000 and the local people said that while the *panchayat* only got Rs. 50,000, the balance of Rs. 7,950,000 was shared by some of the key people, including members of the *panchayat*.

The fallout of this was widespread. When we had a meeting with the members of the women's group in Aklaspur village, a man suddenly came and took over the running of the meeting. He thought I had come as part of the IDS team to encourage the women to stand up against the *panchayat*, and he started shouting and saying that there was no point in women coming together and fighting the *panchayat* because they had seen what had happened in Medleri. The *sarpanch* was supreme, he said, and if the women opposed the village leaders, they would be dealt with. The man was finally thrown out of the meeting. The women were really upset and asked me what they should do. They said that the contractors had threatened to rape and beat up the women if they tried to prevent the eucalyptus from being cut.

Privatisation of the Commons

Apart from the threat from industry that many students of the subject rightly perceive, there are other problems with developing common property resources, especially when the project does not involve all the people in the village. N.S. Jodha (1990: 9–10) collected data on common property resources (CPRs) from 82 villages in 21 districts of India. His study showed that large-scale privatisation of CPRs has led to a decline in their availability in all the study areas. According to him, this has been directly a result of the land distribution policies of the government. Since the 1950s to date, the state governments have been giving CPRs under various anti-poverty programmes because they were not able to enforce land ceiling laws and redistribute excess land held by big landlords. But these programmes have not necessarily benefited the poor who have undoubtedly received some land. But the proportion of land received by the poorest was much lower than what other social groups were able to get. The poor barely received more than one hectare per household on an average while 'others' received between two and three hectares. A detailed enquiry revealed that the poor were dispossessed of 23 to 45 per cent of the land thus received. As Jodha (1990:10) writes: 'It is quite doubtful whether the poor people's collective loss through reduced CPRs has been compensated by their individual ownership of the erstwhile CPR lands.'

Parmesh Shah and Andrew Weir, analysing the experience of several NGOs in western India, are concerned about the long-term impact of these short-term interventions. They have this to say:

> The long-term problems of ensuring equity remain unresolved .... Equity cannot be forced and conflicts are bound to arise. NGOs can only attempt to establish sensitive and democratic village institutions and advise and encourage their development. However, in many cases, the NGO by taking a target population approach (e.g., the poor) alienates the population affected by their intervention. (1987: 18–19)

N.C. Saxena is also concerned about the privatisation of the commons. In a paper based on the Uttar Pradesh experience he writes: '... creating rights in favour of some on resources on which the entire village is dependent would not appear fair and may lead to social tensions, even within the category of the poor.' (ibid., 1987: 15)

Anil Agarwal and Sunita Narain are also against privatisation. According to them:

> In a densely populated country like India the privatisation approach effectively reduces access to the commons for a large number of village people, while giving control to a few members of the community. There is not enough land to benefit all the poor people on a private basis. There will always be some poor people who will be left out. Given that these lands are vital for the survival of the poor, landless and marginal farmers who use the lands for fodder, fuel, grazing and other subsistence needs, they will be adversely affected by the reduction of common lands. (ibid., 1989: 15).

Agarwal and Narain give examples from Rajasthan where beneficiary families have even been excommunicated by the village because of this privatisation (ibid.).[17]

IDS has faced a similar experience with a community fodder farm they have developed at Aryamallepur village, Karnataka. In 1984, the government gave 25 acres of revenue wasteland to a cooperative society on a 15-year lease to develop a fodder

[17] See also Shiva (1988: 79–95).

farm because the villagers had complained of acute fodder shortages. The village women were particularly anxious because they had organised a dairy cooperative. The responsibility for implementing the project was with IDS who organised 13 people from landless families of whom eight were women and five were men. They belonged to several communities—Lingayat, Maratha, Kuruba, Naik and Harijan.

Government and other donor funds were used to prepare the plantation, dig a borewell, make a workshed and a shed for dairy animals. The idea was that with bank loans, the people could buy milch animals and become part of the dairy cooperative but this did not happen as they had taken other bank loans earlier which had not been repaid. IDS tried over a period of four years to make the people feel that it was their farm and not an IDS enterprise.

The plan for the community fodder farm was that villagers would purchase fodder from the farm. Since the members of the society were getting wages, the sale proceeds would be used for generating a community fund to be used after the external resources were over to ensure that the fodder farm could run independently.

A number of things went wrong with this plan. For one thing, the villagers resented the fact that a small group of landless people were getting wages for the work they were doing so they retaliated by not buying the fodder. Some of them broke the fence at night and let in their cattle. Second, other landless villagers wanted to participate in the project, but to have accepted more people would have meant that the project would not have been economically viable. The crisis reached a head during the drought years of 1987–88 when there was demand for work even from small and marginal farmers. During this period, project beneficiaries were the only ones to get full work at minimum wages. Third, while these 13 people had initially agreed to reduce the number of persons working full-time on the plantation to six or seven and to then work in shifts, when this did happen, they did not like it at all. Although the IDS had made it clear from the start that they would withdraw from the project at the end of 1989 and the wages would have to be paid from the profits, and the members had agreed, when IDS withdrew and the members had to run the farm on their own, there was a lot of bitterness.

Analysing their experience with the farm, Ms. Hiremath, the Project Coordinator said:

> The wasteland project which was meant to help the village has virtually split the village. This group of landless people are resented by not only the higher caste and land-owning families, but also by others of their own caste and kin.

She explained further:

> The community fodder farm was seen as privatising the commons. When the village did not have water, the farm had a borewell. Although the·‸ were 250 acres of *gomaal* land available to the village, now there is hardly any left. Our hunch is that people started encroaching soon after we got this 25 acres thinking that if they do not take it, the government will give it away to other social groups. As a result, there is hardly any *gomaal* land left for free grazing.

There was also a feeling that now that the land was fertile again, there were *panchayat* members who had their eyes on it and they were deliberately creating mischief to try to break up the group. At the meeting that I attended, none of the members of the cooperative saw themselves as belonging to the cooperative as they had always received wages and saw themselves as *coolies*. They were waiting for the IDS to hand over the farm to the *panchayat* so that they could continue to work on the farm as *coolies*. They did not see how this collective asset could be theirs as their notion of an asset was one on which they had *patta* rights.

Despite the views of students of the subject (cited above), there are several examples from the field to suggest that dissension is not inevitable when a sub-group like women, landless people, Scheduled Castes or Tribes are targeted as beneficiaries in a wasteland project. PEDO in Rajasthan, SARTHI in Gujarat, CWDS in West Bengal and DGSM and JSS in Uttar Pradesh have successfully organised their projects around poor women. As pointed out earlier, dissension is less likely to occur in single-caste or single-tribe villages, where there is a high rate of male out-migration or where the projects are located in geographically more remote areas. If a project does not define its beneficiaries, it is not certain that its benefits will flow to the disadvantaged in the population (Skutsch, 1989: 61). General developmental experience suggests that, most likely, it will be the better off who will prosper. Can we realistically expect that, in the absence of an identified set of beneficiaries, the poor will get their due share of the project's benefits? This question merits further debate and consideration.

Women's Participation: Some Facets

What is the justification for targeting women as an identifiable group to benefit from a development programme? If the theology of women's liberation has taught us gender-sensitivity, the reality of development programmes has pointed to the need for gender specificity. A whole range of well-intentioned generalised development efforts have been found to leave out of their scope and impact vulnerable groups like the Scheduled Castes and Tribes, small and marginal farmers, landless labourers and so on. Given the pattern of distribution of assets and power in society and family, the fruits of development tend to be picked by the better off social and economic groups in society. It is now widely acknowledged that women as a sub-group deserve special assistance to improve their position in the family and in the economy (Behal, 1990). While recognising that environmental degradation increases the burden of women in their quest for fuel and fodder, I do believe that it is not necessary to mythologise women's relations with nature in order to support the case for women's participation in wasteland development.

It will not be out of place to mention here that the way a project on women and wasteland development is conceived and run depends importantly upon whether one's emphasis is upon women or wasteland development. For those who look at things with ecologically tinted glasses, the regeneration of the wasteland is the first and most important objective. In such a scheme, not much attention is paid to income-generation for women. As a result, it is questionable whether the interest of women participating in the project can be sustained over the long period of gestation that plantation necessarily involves. For those who are concerned about wasteland development but equally focus on the role of women, their projects exhibit more detailed planning and awareness of the longer term. In Rajasthan, one agency sought to establish forward linkages by concentrating efforts to grow fodder species in villages located along the milk routes of dairy cooperatives, thus projecting and integrating current wasteland efforts with future economic opportunities. In another instance from West Bengal, CWDS sought to encourage *arjun* and *asan* plantations with an eye to the commercial potential of rearing *tassar* cocoons for the market. In balance, it is clear that the participation of women can better be secured where income-generation for them is seen as an integral part of the larger goals of the project.

In different states of India, governments through their social forestry programmes encouraged plantations of monocultures like eucalyptus and pine. There have been criticisms of this policy from different perspectives: sometimes, the selection of species like eucalyptus was questioned on the grounds of the damage resulting to the ecosystem; others saw such moves as a capitulation to the needs of commerce and industry rather than as a response to the needs of people for fodder and fuel. As a consequence of government policy, there has been a backlash amongst environmentalists against the very notion of planting one or two species only. One of the criticisms of the CWDS's project in West Bengal has been precisely that the cultivation only of *arjun* and *asan* was not ecologically desirable and did not address fuel and fodder needs. Looked at from the micro-context, however, the tribals involved with the project neither have a tradition of livestock-rearing nor do they see the supply of fuel as a high priority as it is locally available. In such circumstances, it would seem rational to develop wasteland with a view to providing supplementary income through *tassar* cocoon-rearing. From the point of view of local tribals, faced as they are with a chronic lack of employment, the income-earning opportunities offered by cocoon-rearing are clearly more attractive than the prospects of having their fuel needs satisfied with a long-gestation plantation. The point that emerges from this example is the importance of regional variations and local priorities.

Sometimes, there is a conflict between the needs of women and the protection of the environment. A growing concern amongst experts on the environmental issue is that people should reduce the number of animals they have or improve the quality of livestock so that there is less pressure on natural resources. Concretely, one implication is that animals have to be stall-fed, a task invariably done by women. This is necessarily time-consuming and takes away valuable time from daily wage work (ILO/ARTEP, 1988: 48; Molnar and Schreiber, 1989: 11).

Conversely, women prefer goat-rearing, particularly in the more drought-prone areas. Goats are the biggest and most valuable asset of poor, landless families but goats are environmentally unfriendly. N.S. Jodha (1990: 17) states that one of the main effects of the reduction of common property resources (CPR) is the substantial increase in the number of goats and sheep. Thus the sheep and goats, often accused of destroying vegetation in the CPR, seem to have become more important following the degradation of CPR, rather than the other way round. He writes about other micro-level studies which have revealed the same trends.

In the projects I visited and studied, the nature and extent of women's participation varied widely. In some cases, their participation did not amount to more than being paid wages for their labour on plantations and they were not consulted in the design, conception or implementation of the project. In almost all the projects, women were actively involved in nursery-raising. In some projects that I visited, the plantations seemed excellent but pointed enquiries about how far women truly participated in the exercise revealed that all major decisions were taken by personnel technically qualified in agriculture. Project staff expressed the opinion that ordinary people, particularly women, did not have the technical expertise required for the task at hand. There was another project in which discussion suggested that women were deeply involved in the running of the project but the plantation itself was poor. Further analysis brought out the fact that intra-group conflicts among the women had led to some people wilfully grazing their cattle on the plantation to its detriment. But the poor end-result could not be attributed to the lack of technical understanding about agricultural techniques. Along the spectrum of NGO efforts, one noticed different mixes of professionalisation on the one hand and women's participation on the other. Perhaps what is needed is an approach that would harness the skills of agricultural scientists in the service of the needs of the people. This would probably require some reorientation on the part of those technically qualified in agriculture, marketing and related activities so that genuine dialogue could ensue between specialists and project beneficiaries. On the other side, perhaps some NGOs need to display greater respect for modern management techniques without sacrificing their commitment to the involvement of ordinary people in designing and executing wasteland projects. One should neither make a fetish of what is scientific or technical nor should one ignore it in the name of traditional knowledge of communities.

In Chamoli district, both Anand Singh Bisht and Chandi Prasad Bhatt had been closely involved with relief work after the Alaknanda floods in 1970 and they came to be widely known in the area. They started mobilising women and helping them make the connection between deforestation, landslides and floods. At first, there was little response from the women. Both Bhatt and Bisht said they started their work by going on *padyatras* to villages. At night, they would halt in a village and the villagers would look after them. For many years, they discussed the problem of environment and its impact on women. Gradually, women started attending their meetings. This led to the idea of setting up or reactivating *mahila mangal dals*.

Both JSS and DGSM follow the same approach as far as plantation work is concerned. They hold a series of meetings in villages where people have shown interest in protecting the environment. In these meetings, the first attempt is to mobilise the women on issues which are of primary concern to them. Thus, for example, when women say that their crops are being attacked by wild boars and bears, the discussion centres around possible solutions. In most villages, the women have wanted boundary walls around their own agricultural lands. It is then suggested to them that if they plant trees on the wasteland surrounding their fields, it would act as an additional buffer. This idea is easily acceptable to the women and this is how they initially get involved with plantation work.

What was truly exciting about the women I met through these two organisations was their strength and confidence in their ability to handle any adversity on their own. It was not just a question of one or two women but of almost every woman I met, old and young alike, feeling that they were ready to deal with anyone and anything. All the activists—Bisht, Bhatt, Murari Lal and Chander

Singh Rana—were catalysts and facilitators but did not want to take any credit for the work being done.

One of the factors responsible for a high rate of women's participation in some of the projects described is the male out-migration that has taken place. The lack of employment opportunities in the UP hills, in Rajasthan and in parts of Gujarat lead a significant proportion of the male population to migrate to the plains or towns in search of work. This migration may be either seasonal or permanent but in either case women, children and the old are left behind. Many agencies were in any case gender-sensitive but their task was made simpler by the sheer preponderance of women.

It was also observed that there was a higher participation rate amongst women in those project areas where men had other local employment opportunities. Gunilla Olsson (1988) reports that when work was available at higher wages with the Forest Department, the men would tend to be employed there and women would come in large numbers to work on the plantations. In Rajasthan and in Uttar Pradesh, it was noticed that wherever there was competing and better wage employment available than what was offered on the plantation, it was always the men who seized those opportunities. But in the absence of such alternatives, even the men would come forward to work on the plantations. The dynamics of the social structure ensured that better-paying jobs on the plantations would invariably be done by men. This was sometimes rationalised by the agency on the grounds that the tasks to be performed by men were harder and required greater strength or skills.

Conclusions

When one reviews the experience of these different groups, some broad conclusions emerge. Many of the groups visited began their work on private lands pooled together. This had several advantages: for one thing, there was hardly any scope for dispute or conflict over the right to usufruct and for another, all the technical difficulties involved in the regeneration of wasteland could first be faced and dealt with on private plots. The experience thus gained would be of value when *panchayat* or revenue lands would later be taken up. The question of conflict over sharing the usufruct when common lands are developed is of central importance. In many projects visited, efforts were made to evolve a consensus on this vexed issue and often such consensus would be recorded in writing. Even if, before being developed, common lands do not excite passion because of their poor quality, it is inevitably the case that after development, rivalries erupt unless care has been taken in advance to determine the distribution of produce. Given the place of land in the agricultural economy and the complex web of customary rights, it is not surprising to find that even carefully hammered-out agreements do not provide a guarantee of peaceful sharing in the post-development phase when the true potential of the land becomes visible.

Another generalisation that could be hazarded is that wherever agencies have tried to assess the felt needs of the beneficiaries, it has inevitably resulted in genuine and heartfelt participation of the people in the implementation of the programme. One should not go blinkered into the field assuming as an axiom that fuel and fodder are the most important and immediate requirements. That may well be so in a number of cases but regional variations and local priorities deserve to be carefully taken into account and particular needs established. It is quite possible that even in situations where people have significant problems of fuel and fodder they may still want plantations that augment their incomes. In one example from Orissa, a voluntary agency was forced to go in for fodder and fuel varieties because those were the terms of the grant from the funding agency. The local tribal community did not have problems of either fuel or fodder but were interested in growing fruit trees and bamboo for the monetary benefit they would bring. In contrast, another group in Orissa—PREM—consciously tried to respond to their perception of what tribal needs were and, in the particular case, these were thought to be related to better income-earning opportunities rather than fuel and fodder. Hence they went on to plant fruit trees.

Many of these agencies surveyed in this paper did not begin with women and wasteland projects. They established a presence in the area, engaged in dialogue with local communities and sought to understand their priorities. The JSS and the DGSM are good examples of this kind of group. As mentioned earlier, in some villages women complained that their crops were being destroyed by bears and wild boars. So the agencies suggested that boundary walls be built and trees planted on the wasteland surrounding the croplands to act as buffers. Once the immediate needs of the people were taken care of, wasteland work was taken up very enthusiastically. In the process, the credibility of the agency was underlined and people responded favourably to their environmental overtures. It would therefore seem to follow that, when government or funding agencies are considering proposals for wasteland development through NGOs, they should consider the extent to which the NGO has struck roots in the community concerned. Given the long gestation period that wasteland development necessarily entails, it would seem important to involve women in supplementary income-generation schemes such as dairying, goat-rearing, *tassar* cocoon-rearing, fruit processing, etc. This list is obviously only illustrative and the type of schemes formulated will have to be based on local resources, skills and market situations.

The integration of such schemes into a project on wasteland development is more likely to nurture and sustain interest in wasteland development itself. We noticed in the Uttar Pradesh hill villages, where DGSM and JSS have actively involved women in wasteland development and have made them aware of its ecological and social implications, that women in the area are now asking for something over and above wasteland development itself.

An imaginative effort to link wasteland development with the processing and marketing of fruit grown on that land faltered for extraneous reasons. An arrangement was secured with a public sector agency to purchase the oranges with a view to marketing them in the plains. The agency offered so low a purchase price that even costs of cultivation were not recovered and the economic viability of the project was endangered. Attempts to set up a small processing facility to convert the produce into juice were unsuccessful for lack of permission from the relevant authorities. This story signals the importance of professional market support in setting up backward and forward linkages. Not all NGOs may have relevantly qualified personnel or access to such professional support.

It is clear that there are many factors that need to be taken into account in planning a programme involving women in wasteland development. There are also many factors that facilitate the process of women's participation. One of the factors is homogeneity of the villages both in terms of caste and class. The most successful projects are those which have not had to deal with inter-caste rivalries and tensions. When people belong to the same social class, there is much less dependence upon each other in a hierarchical patron–client relationship. One could hardly expect equal participation in a project where the target group consisted both of large landowners and the landless. Given the highly stratified nature of Indian society and the fact that at the village level in particular, social, economic and political power tends to be concentrated in the same hands (Saxena and Gopalan, 1989: 24; Verma, 1988: 20; ILO/ARTEP, 1988: 46), poor and landless, Harijan and tribal villagers are unlikely to voice their opinion or participate in the activities of the project when there are high caste, land-owning people present—except as wage workers. Even when the project is for women only, it is difficult to bring these different castes and classes together. When NGOs are conscious about caste and class differences and have a well-established presence in the area, it is possible for them to work towards breaking down these barriers as JSS and DGSM have tried to do. But this is more easily possible in areas like the hills than in the plains where there would be other countervailing forces also present.

We have tried to argue that while there is a strong case for targeting women as a sub-group in wasteland development, that case should not rest upon an alleged link between women and nature; rather, it should be buttressed by the widely recognised fact that there is pervasive inequality between genders in most of our social groupings. Examples have been cited where it was found that the true opposition lay not between poor women and poor men but rather it lies between the needs of the poor as such and the more powerful but competing interests of commerce and industry. The desirability of privatising common lands was raised in a dramatic fashion when vast tracts of common lands were handed over for industrial needs despite an endemic shortage of fuel and fodder in Karnataka.

We have noted earlier how the nature and extent of women's participation in the wasteland projects visited depend in part upon the orientation of the NGO. This apart, male out-migration, which occurs in large parts of Rajasthan and in the hills of Uttar Pradesh, is one of the objective factors that predisposes towards greater women's participation. It is not only the arithmetic of gender presence that accounts for increased participation of women but also the fact that in the absence of their men, women are forced to play a greater role in decision-making than would probably otherwise have been the case. But it is clear that the orientation of the NGO is the crucial determinant of the degree of involvement of women. Some groups view women only as a subset of the poor and their participation in wasteland projects is not explicitly sought though it might well take place in the natural course. The level of participation was also found to vary significantly with the project concerned: sometimes, participation might simply mean wage employment; at other times, women would be involved in nursery-raising, but not in plantation work; in some cases, conscious attempts would be made to secure their participation through all stages of conceiving and executing the projects. We have also noted how some adjustment is necessary as between the requirements of agricultural technology and the participation of the beneficiaries. It is necessary to arrive at a *via media* that pays obeisance both to science and the people. A practical compromise between the purely technological and the wholly participatory seems to be called for.

We have tried in this paper to describe the differing experiences of various voluntary agencies in involving women with the development of wastelands. Their varied experiences are not only important in themselves but also for the more generalised issues that are thrown up from micro-level contexts. The conflicts between groups and subgroups, different castes, the better-off and the poor and between the poor and vested interests need to be recognised by policy-makers, funding agencies and NGOs themselves. Then the search for solutions that are harmonious would perhaps become easier.

REFERENCES

Agarwal, Anil (1987) 'Between need and greed—the wasting of India: The greening of India.' In Anil Agarwal, Darryl D'Monte and Ujwala Samarth (Eds.) *The Fight for Survival*. New Delhi: Centre for Science and Environment.

——— (1988) 'Gandhi's ghost saves the Himalayan trees.' In Kunwar Shishupal Singh (Ed.) *The Chipko experience: Hugging the Himalayas*. Gopeshwar: Dasholi Gram Swarajya Mandal, pp. 17–20.

Agarwal, Anil and Sunita Narain (1989) *Towards green villages: A strategy for environmentally sound and participatory rural development*. New Delhi: Centre for Science and Environment.

——— (1990) *Strategies for the involvement of the landless and women in afforestation: Five case studies from India. A technical cooperation report*. World Employment Programme. Geneva: International Labour Organisation.

Behal, Monisha (1990) *Social forestry in Bihar: Special reference to NGOs and involvement of women in the project area*. Indo–Swedish Forestry Coordination Programme. ISO/Swed. Forest Consultants in Forestry, SIDA.

Bhatt, Chandi Prasad (1990) 'Forest conservation based on people's power and people's aspirations.' In Ramesh Chandra (Ed.) *Role of panchayat in social forestry and wasteland development*. New Delhi: National Mission on Wastelands Development, Ministry of Environment and Forests, pp. 10–11.

Bhattacharjea, Ajit (1987a) 'Reverse development—I. Satyagraha in Kusnur.' *Deccan Herald*, 19 November.

———(1987b) 'Reverse Development—II. State-aided exploitation.' *Deccan Herald*, 20 November.

——— (1988a) 'Kusnur struggle phase two—I. KPL strikes back.' *Deccan Herald*, 5 May.

——— (1988b) 'Kusnur struggle phase two—II. A corrupted system. Significant success.' *Deccan Herald*, 15 June.

Chambers, Robert and Melissa Leach (1987) 'Trees to meet contingencies: Savings and security for the rural poor.' *ODI Social Forestry Network*, Network Paper 5a, October, pp. 1–29.

Chambers, Robert, N.C. Saxena and Tushaar Shah (1989) *To the hands of the poor: Water and trees*. New Delhi: Oxford and IBH Publishing Co. Pvt. Ltd.

Chandra, Ramesh (1990) 'Role of panchayats as extension agents of wasteland development.' In Ramesh Chandra (Ed.) *Role of panchayat in social forestry and wasteland development*, New Delhi: National Mission on Wastelands Development, Ministry of Environment and Forests.

Chowdhry, Kamla (1983) *Social forestry and the rural poor*. Discussion Paper No. 10, Ford Foundation.

Dankelman, I. et al. (1989) 'Women and forest: Fuel, food and fodder.' In Brigitte van der Borg (Ed.), *Women's role in forest resource management: A reader*. Regional Wood Energy Development Programme in Asia, Project Working Paper No. 1. Bangkok: Food and Agricultural Organisation of the United Nations, December, pp. 9–20.

Dhagamwar, Vasudha and Enakshi Ganguly Thukral (1990) 'Legal, illegal and social-legal problems of women in wasteland development.' Paper prepared for the ILO Workshop, *Women and Wasteland Development*, New Delhi, 9–11 January 1991.

Fernandes, Walter and Geeta Menon (1987) *Tribal women and forest economy:*

Deforestation, exploitation and status change. Tribes of India. Series 1. New Delhi: Indian Social Institute.

Fernandes, Walter, Geeta Menon and Philip Viegas (1988) *Forests, environment and tribal economy: Deforestation, impoverishment and marginalisation in Orissa*. Tribes of India. Series 2. New Delhi: Indian Social Institute.

Fortmann, L. and D. Rocheleau (1989) 'Why agroforestry needs women: Four myths and a case study.' In Brigitte van der Borg (Ed.) *Women's role in forest resource management: A reader*. Regional Wood Energy Development Programme in Asia, Project Working Paper No. 1. Bangkok: Food and Agricultural Organisation of the United Nations, pp. 21–32.

Freerk, Wiersum K. (1986) *Social forestry and agroforestry in India*. Report of the Mission of the Government of the Netherlands on the identification of the scope of Forestry Development Cooperation in India, 14 April–2 May.

Gadgil, Madhav (1989) *Deforestation: Problems and prospects*. Foundation Day Lecture, 12 May, New Delhi: Society for Promotion of Wastelands Development, pp. 1–44.

GOI (1988) *National Forest Policy 1988*. New Delhi: Government of India, Ministry of Environment and Forests, pp. 1–6.

Guha, Ramachandra (1989) *The unquiet woods: Ecological change and peasant resistance in the Himalayas*. Delhi: Oxford University Press.

ILO (1988) *The Bankura story: Rural women organise for change*. New Delhi: International Labour Organisation.

ILO/ARTEP (1988) *Employment and income generation through social forestry in India: Review of issues and evidence* . Report prepared for the Government of India and the Swedish International Development Authority (SIDA), New Delhi.

Jha, Shekhar Kumar (1988) 'The tendu leaf tangle'. *Commerce,* 30 November–6 December, pp. 17–18.

Jhala, B.K. (1990) 'Village woodlots in Gujarat'. In Ramesh Chandra (Ed.), *Role of panchayat in social forestry and wasteland development*. op. cit., pp. 45–53.

Jodha, N.S. (1990) *Rural common property resources: Contributions and gains*. New Delhi: Society for Promotion of Wasteland Development. Foundation Day Lecture, 16 May, pp. 1–24.

Kaur, Ravinder (1988) 'Women's role in forestry in India.' Consultant Report for the India Country Paper. Women in Development Study, Washington, D.C., World Bank, PHRWD, September.

Khator, Renu (1989) *Forests: The people and the government. An analysis of the forest policy performance in India* . New Delhi: National Book Organisation.

Khetan, Neelima (1989) *Wasteland development: A review (1986–89)*. Udaipur: Seva Mandir.

Longhurst, Richard (1987) 'Household food security, tree planting and the poor: The case of Gujarat.' *ODI Social Forestry Network*, Network Paper 5d, October.

Malhotra, K.C. and Mark Poffenberger (Eds.) (1989) *Forest regeneration through community protection: The West Bengal experience*. Proceedings of the Working Group Meeting on Forest Protection Committees, Calcutta, 21–22 June. Sponsored by West Bengal Forest Department, Indian Institute for

Bio-Social Research and Development, Ramakrishna Mission, Ford Foundation, pp. 1–47.

Mistry, M.D. (1987) *Legislative protections to the women labourers engaged in forest-based occupations. A report.* Sabarkantha: DISHA, pp. 1–54.

Mistry, Madhusudan and Vimala Kharadi (1990) *Cries in the wilderness: A report on the exploitation of tribal women tendu leaf collectors by traders and Gujarat State Forest Development Corporation*. Sabarkantha: DISHA, pp. 1–22.

Molnar, Augusta and Gotz Schreiber (1989) *Women and forestry: Operational issues*. Policy, Planning and Research Working Papers. Women in Development, Population and Human Resources Department, the World Bank, pp. 1–47.

NABARD (1987) *Refinance assistance for development of wastelands*. Unpublished, pp. 1–9.

NCAER (1988) *A review of social forestry projects and programmes in selected states in India*. Report prepared for the Government of India and the Swedish International Development Authority (SIDA). New Delhi: National Council for Applied Economic Research (NCAER).

Olsson, Gunilla (1988) *Employment and income generation in social forestry: A case study from Orissa*. Geneva: International Labour Office.

PEDO (1989) *Summary Report*. Jan Shiksha Avam Vikas Sangathana (People's Education and Development Organisation). Mada, Dungarpur District, Rajasthan.

Poffenberger, Mark (1989) *Land and forest management in South Asia*. A Ford Foundation Strategy Statement. (Draft), June, pp. 1–42.

———(1990) *Joint management for forest lands: Experiences from South Asia*. A Ford Foundation Program Statement, pp. 1–49.

PRIA (1984) *Women and wasteland development* . New Delhi: Society for Participatory Research in Asia.

Saxena, N.C. (1987) 'Commons, trees and the poor in the Uttar Pradesh hills.' *ODI Social Forestry Network* . Network Paper 5f, October.

———(1989) 'Women in forestry.' Paper presented at the *Workshop on involvement of women in social forestry*. National Wastelands Development Board, India International Centre, 10–11 February, New Delhi, pp. 15–23.

Saxena, K.B. and Sarala Gopalan (1989) 'Access of women to land for social forestry: Talking points.' Paper presented at the *Workshop on involvement of women in social forestry*. National Wastelands Development Board, India International Centre, 10–11 February, New Delhi, p. 24.

Seva Mandir (1989) *Annual Report 1988–89*. Udaipur.

Shah, Parmesh and Andrew Weir (1987) 'Approaches to social forestry in Western India: Some aspects of NGO experience'. *ODI Social Forestry Network*. Network Paper 5b, October.

Shah, S.A. (1990) 'Socio-economic strategies to save tropical forests of India.' *Wastelands News*, Vol. VI, No.2, November 1990–January 1991, pp. 2–14.

Sharma, Kalpana (1987) 'Rural poor or industry? Who should benefit from common lands forestry?' *Indian Express,* 19 May.

Shah, Tushaar (1987) 'Gains from social forestry: Lessons from West Bengal.' *ODI Social Forestry Network*. Network Paper 5e, October, pp. 1–20.

Shepard, Mark (1988) 'Chipko: North India's tree huggers.' In Kunwar Shishupal Singh (Ed.), *The Chipko experience: Hugging the Himalayas*. Gopeshwar: Dasholi Gram Swarajya Mandal, pp. 24–35.

Shepard, Gill (1986). 'Forest policies, forest politics.' *ODI Social Forestry Network*. Network Paper 3a, October, pp. 1–31.

Shiva, Vandana (1988) *Staying alive: Women, ecology and survival in India*. New Delhi: Kali for Women.

Shiva, V., H.C. Sharatchandra and J. Bandyopadhyay (1981) *The social, ecological and economic impact of social forestry in Kolar* (mimeo). Bangalore: Indian Institute of Management.

Singh, N.K. (1989) 'The tendu leaf issue troubles the Government.' *India Today*, 15 April, pp. 63, 65.

Skutsch, Margaret, M. (1989) 'Women's access in social forestry.' Guide to literature. In Brigitte van der Borg (Ed.), *Women's role in forest resource management: A reader* . Regional Wood Energy Development Programme in Asia, Project Working Paper No. 1. Bangkok: Food and Agricultural Organisation of the United Nations, December, pp. 60–78.

SPWD (1984) *Hill resource development and community management*. Sukhomajri and Dasholi Gram Swarajya Mandal. New Delhi, August, pp. 1–14.

——— (1988) *Proceedings. National Workshop on Small Scale Watershed Development*. Suraj Kund, 30 October to 1 November. New Delhi: Society for Promotion of Wastelands Development.

UVM (1989) *Ubeshwar Vikas Mandal — Dhar. Annual Report*. 1988–89, Udaipur.

Verma, D.P.S. (1988) 'Women and forest nurseries in Ganeshpura village.' *The International Tree Crops Journal*, 5, pp. 19–29.

Wasteland News. Vol. VI, No. 2, November 90–January 91.

Yadav, Haridai Ram (1986) *Genesis and utilisation of wastelands: A case study of Sultanpur district*. New Delhi: Concept Publishing Company.

Zahedi, Mehboob (1990). 'Role of panchayat in social forestry and wasteland development with special reference to Burdwan district (West Bengal).' In Ramesh Chandra (Ed.), *Role of panchayat in social forestry and wasteland development*. New Delhi: National Mission on Wastelands Development, Ministry of Environment and Forests, pp. 67–70.

7 *Women and Wasteland Development in India: Policy Issues*

N.C. SAXENA

Introduction

In this paper the term 'wastelands' is used loosely to apply to all uncultivated lands which are capable of, or are, supporting vegetation.

Indian forests are capable of producing at least two cubic metres of wood per hectare per year (FSI, 1988) as against the present productivity of 0.47 cubic metres. For the other category of degraded land, the Agriculture Finance Corporation (a Government of India Undertaking) estimated that an investment of Rs. 9000 to Rs. 15,000 per hectare (including overheads) in irrigation, soil improvement and protection would increase the productivity of vast blocks of village wastelands from almost nil at present to four to eight cubic metres per hectare per year (Saxena, 1989b). Thus wastelands have immense biological potential which is sought to be realised through several social forestry schemes. However, the experience of the last 10 years shows that the benefit to women from these schemes has been, at best, marginal. This paper discusses the policy issues which would help women through forest and wasteland regeneration.

There are three major types of land—forest, revenue and private. Women are most dependent on forest lands, where they are gatherers of forest produce for subsistence and sale. They are also employed by the Forest Department and contractors to work as unskilled labour. They have similar roles as collectors and as wage-employees on common and revenue lands, though to a lesser extent, as these lands are more degraded. In community forestry programmes, women are supposed to participate in the management of afforested areas. Women are also involved as producers in farm forestry programmes. Thus, women have four distinct occupational roles in forestry—gathering, wage-employment, management and production.

The different roles that women have on different types of land—forest, revenue and private—are well-documented. However, some of the major problems that women face as gatherers of forest produce and as wage-employees of the Forest Department and contractors, need to be discussed in more detail. Similarly, their roles in community forestry programmes and as producers in farm forestry need to be highlighted.

Women as Gatherers

Main Problems

Women face three main problems with respect to gathering forest produce. The reason for this is that their rights of collection are not publicised enough. Many State Governments have nationalised minor forest products (MFPs) in the interest of collecting revenue. And opportunities for self-employment through collection of forest products are on the decline.

Lack of information

Often women are not aware of what they can collect from forest lands. This leads to harassment. A Government Commission on Women interviewed Dapubai (a tribal woman in Udaipur) who got only Rs. 7 for 10 kilograms of gum. It took her 10–12 days to collect although its market price was Rs. 250 (GOI, 1988: v). When asked to comment on the low price, she said, 'How can I demand a higher price? The trader's man threatens to report me to the Forest authorities for entering the forest area. Then we will get nothing.'

Informing women of their rights is not considered politically beneficial, whereas leaving them to the mercy of the low-paid forest staff is perceived as politically neutral. Such is the irony of the Indian political system!

Nationalisation of the MFPs

Before nationalisation, the gatherers could sell forest produce to anybody, but under the new system it has to be sold to the Forest Department only. In almost all cases the Department has appointed agents formally or informally (GOI, 1987). This has put the gatherers at the mercy of two different sets of people, the contractor as well as the government department and payment gets routed through both of them (FAO, 1989: 70). For instance, private trade in *sal* seeds is illegal in Madhya Pradesh, but shopkeepers manage to exchange it with tribals for daily necessities at a low price. They then sell it to government bodies, thus defeating the very purpose of nationalisation (GOI, 1988a).

A Government Commission was told by the tribal women in Madhya Pradesh that when they walked a long distance to the office of the Forest Department to sell their produce, they often found it closed, or were told to come the following day. This forced them to sell to the traders at only 20 per cent of the government price (GOI, 1988: xxv).

After nationalisation, the collection of *sal* seeds fell in India from 200,000 tonnes in 1979 to only 60,000 tonnes in 1987. Production of lac, another nationalised item, has gone down from an average of 32,000 tonnes per year during 1961–70 to 16,000 tonnes during 1981–86 (GOI, 1988).

Nationalisation reduces the number of legal buyers, chokes the free flow of goods and delays payment to the gatherers, as government agencies find it difficult to make prompt payment. This results in contractors entering through the back door but they must now operate with higher margins required to cover uncertain and delayed payments by government agencies as well as to make the police and other authorities ignore their illegal activities. All this reduces women's collection and incomes.

Declining self-employment

Vast potential exists for quadrupling employment opportunities through the collection of MFPs from 0.8 million person years to

3.3 million person years (GOI, 1976). But in reality, although government revenues from MFPs have increased substantially in the past two decades, total employment and incomes have at best remained stagnant, and have quite often fallen; and the quantities collected and the payment to collectors have rarely risen.

In Madhya Pradesh, the production of *tendu* leaves fell from 5.10 million bags in 1981–82 to 3.9 million bags in 1985–86 (GOMP, 1987: 33). In Orissa, despite nationalisation, production of *tendu* leaves over a longer period has remained stagnant at around 36,000 metric tonnes a year in the past two decades (Gupta and Guleria, 1982; GOO, 1988: 8).

From the point of view of poor women, it is not just volume but remuneration that matters. After meeting all expenses and profits they should be paid Rs. 1.31 a kilogram for collection of *sal* seeds in Madhya Pradesh, but they get only Rs. 0.55 a kilogram (GOI, 1988: 55). In Madhya Pradesh and Orissa, the price which tribals get for tamarind is as low as Rs. 1.40 to Rs. 1.75 per kilogram whereas it sells for Rs. 9 to Rs. 11 per kilogram in the Bombay market. Sometimes the traders follow a barter system by offering one kilogram of sweet potato costing less than a rupee for one kilogram of tamarind (GOI, 1988: 40), which they sell at five times the price in other markets.

Women as Wage-Employees

Women are preferred by the forestry staff and contractors for certain forestry operations like nursery work and *tendu* leaf collection. However, they often get lower wages than men for similar work, are not paid regularly and are subjected to harassment if they complain (CIDA, 1988).

An ILO study (Olsson, 1988) of the Social Forestry Programme of Orissa observed that nowhere in the Appraised Project Document was there any mention of the working conditions of women—the fact that they got no benefit of labour laws and that no safety or health measures were being undertaken, whereas work being performed outdoors, under exposure to changing weather, required heavy physical effort, sometimes in difficult terrain, and away from their homes. The Village Forest Committee was generally not involved in payment of wages, which in any case were lower than the minimum prescribed. Wages, as stated by the Forest Department, were higher than what women actually got. Wages paid to men were higher than wages paid to women but women were not supposed to reveal this secret to outsiders. When asked why they didn't complain, they replied that they were afraid of the consequences and said: 'What is the use?'

It has been estimated that the total wage-employment for women in the collection of forest produce is as high as 300 million women days (Pant, 1980). Yet hardly any rules exist for regulating their working hours, safety precautions, provision of latrines, job recruitment, leave and other benefits, training policies, productivity-linked bonus, compulsory insurance against accidents, shelters, civil amenities, creches, arrangements for the care of children and infants and medical care. The same is true of forestry work undertaken under the Jawahar Rozgar Yojna (JRY), the National Rural Employment Programme (NREP) and the Rural Labour Employment Guarantee Programme (RLEGP) budgets which flow from the Department of Rural Development, which supposedly looks after the interests of the poor.

Half of the block plantation by farmers in recent years has been on previously cropped lands (IIPO, 1988: 61). A similar conclusion was reached by an ILO/ARTEP study which estimated that 50 per cent of the land covered under the farm forestry component was good agricultural land (1988: 17). How does this affect employment?

By planting trees on land previously used for agricultural crops, female labour tends to get displaced (ILO/ARTEP, 1988: 21; Arnold et al., 1988). A study of eucalyptus plantation in Tamil Nadu under the farm forestry programme (Malmer, 1987) on lands which were previously being used for groundnut cultivation has found that instead of women's employment which groundnut cultivation generated, eucalyptus required digging pits and clear-felling, both of which are done by men. Averaged over a rotation cycle of ten years, total employment per hectare per year dropped from 112 to 45. Female employment dropped from 100 days to nil, while male employment rose but only from 12 to 45. Thus, not only was total employment reduced when plantation trees replaced agricultural crops but women were completely thrown out of employment.

Women as Managers in Community Forestry

In community plantations, who gains and who loses has been affected by the choice of species. Fodder is crucial to the economy of Indian villages. Realising this, land laws of most of the states forbid the use of grazing lands for purposes other than producing grasses and fodder. Yet the Forest Department has planted non-browsable species like eucalyptus on such lands in Gujarat (PEO, 1988), Karnataka (Brokensha, 1988) and other States (IIM, 1985; Arnold et al., 1988), thus depriving poor women of an important resource.

Women tend to lose not only fodder but also other benefits from the trees themselves. A World Bank/USAID team, after touring UP, Gujarat, HP and Rajasthan in February/March 1988, found that commercial species planted by the Forest Department on grazing lands tempt the *panchayats* to sell these rather than distribute them in the village (USAID, 1988). In village Medleri, district Dharwar, Karnataka, a local voluntary organisation had to obtain a stay order from the High Court against the auction of eucalyptus to urban contractors (*Indian Express*, 23 April 1988, and *Deccan Herald*, 23 July 1988), pleading for equal distribution to all families within the village.

Women representatives in village councils are often taken for granted or are subject to strong domestic and community pressures to conform to the conventional norms of their societies. In Bihar, a woman was not even aware that she had been elected as a member of the Village Forest Committee (GOB, 1987:13). Shakuntala of village Nagrota Suriyam, Himachal Pradesh, had the following to say about the council of her village:

> There is a village *panchayat*. For the last 40 years, the *pradhan* has been from the same family. Even the *up-pradhan* (vice-chairman of the village of the village-elected council) works for the welfare of affluent people. No one cares for the poor.

> Lila Devi is a member of the *panchayat* from a reserved constituency. She is not literate. She does not attend the *panchayat* meetings. But she is made to sign some papers every now and then. But being a *panch* (member of the council) is a status symbol for her. (Dasgupta and Maiti, 1986: 10).

Thus it appears that if women have other priorities than men in the choice of species, these have little chance of getting articulated (Arnold et al., 1988). Therefore, given the sex-segregated and hierarchical nature of Indian society, separate women's organisations would be needed in rural India for many years until poor rural women acquire the self-confidence to talk in mixed gatherings. Realising this, some Social Forestry Projects provide for the appointment of women extension workers who are to help in approaching and involving women. A study of Madhya Pradesh showed that within the Forest Department there were only nine women extension assistants out of the total of 273 and only 49 forest workers out of a total of 882 in that cadre. At the level of forest extension officers and higher levels, there were no women. But wherever women forest workers were active, village women not only knew about the general issues concerning social forestry programmes but they were more cost-efficient in nursery management. Even among NGOs, those groups which had separate women staff were more successful in organising women's activities (USAID, 1985).

Yet the social forestry projects in India hardly consider the role of women at the project design or implementation stage. Women have remained 'invisible' to planners and government officials and have been 'designed out' of projects by default (Fortmann, 1983; Fortmann and Ricge Keay, 1985; Molnar and Schreiber, 1989).

Women as Producers in Farm Forestry

As regards farm forestry, the benefits to women are constrained by two factors: their own place in the family and the legal position regarding their ownership of private lands. Unfortunately, over the past several decades women have been deprived of their customary rights in management and ownership of land. It appears that the transition in land rights from communal to private ownership has affected women adversely. So long as land was commonly owned, women had a voice in its management but with private ownership of land, their rights have become diluted. The result is that both extension programmes and credit services are now geared to men. This has helped them to get into the modernised sector while women have remained behind in the subsistence sector.

An FAO study of farm forestry in Gujarat (Jain, 1988) observed that all negotiations for selling the eucalyptus polewood are settled by men whereas women continued to gather fuelwood as before. Agarwal (1988) documents that, in most parts of India, women have had no customary land rights and those that existed have been substantially eroded over time, with state policies playing a catalytic role. One hundred and thirty-one out of 145 land-owning communities studied (Agarwal, 1989) had a patrilineal pattern in land inheritance where women did not get any share. Modern legislation has yet to establish full gender equality in law or to permeate practice.

To give an example (Saxena, 1987a), according to Section 171 of the UP Zamindari Abolition Act and Land Reforms Act, 1950, after a landowner's death, his land will devolve to the male issues in equal shares and in case he had no sons, to his widow and widowed mother. A married daughter would be entitled to a share in the absence of the above claimants only when the deceased had no father, unmarried daughter, brother or unmarried sister. One wonders whether such blatantly unfair provisions of the law are not violating the equality provisions of the Indian Constitution.

Women's right of access to land and other material resources is not a legal issue alone. As their control over loans, income and assets is reduced, their access to social resources such as knowledge, power and prestige diminishes. Disparity in gender status gets intensified with the emergence and deepening of other forms of stratification. Subordination and seclusion of women is more noticed in communities where social differentiation and hierarchy based on the ownership pattern or on prestige are more pronounced.

Currently rural women are not even aware of the necessity of getting separate legal rights over land. They are exploited by their husbands and even by their sons but they do not think of themselves as competitors of men. They would generally like to view their husbands as helpers and friends whose good wishes and advice they would like to cherish. They divide men into the neat categories of good husbands and bad husbands, without realising the inherent exploitation in the very institution of patriarchy and property customs (Ellis, 1988: 170). Even where women have legal rights, they often relinquish their claims to parental land in favour of their brothers (Agarwal, 1988). These norms serve as barriers to women's ability to exercise direct control over the land they may inherit in their natal village. Thus, along with initiating the giving of legal rights over land to women, one would have to conscientise them about the existing realities of power inequities within the family.

While it would require a great deal of political courage to upset the existing power relations between the sexes through legal enactments, a beginning can be made by making women owners of land where new tenurial rights are being created, as on wastelands. Such a limited measure is likely to escape male hostility as the distribution of existing wealth is not involved. It may be pointed out here that Section 3 (c) of the Tree *Patta* guidelines, which states that 'minors and persons who are not capable of planting and looking after the plants themselves may be considered,' is capable of being interpreted against the interests of women (Saxena, 1987a).

Even in such States where special provision has been made to confer tenurial rights on women under the *patta* schemes, it is feared that concrete steps to translate such a provision into action have not been taken so far. For instance, out of 57,546 *pattas* given in UP only 2,949 (five per cent) went to women, despite government instructions that 30 per cent of *pattas* should go to women (Tripathi, 1988). There appears to be a general lack of will on the part of field officials to treat women's interests as distinct from the family's interests.

Possible Solutions

As already stated, of the several roles women have in forestry, the most important is as gatherers of forest produce. Women would therefore be greatly benefited if opportunities for collection from forest and public lands are enhanced. This would require:

—a change in the nature of species being planted on public lands from timber to usufruct trees;
—sharing of management and protection with women;
—publicity about women's rights in forest and community lands and;
—a change in the Forest Policy to facilitate the above.

We discuss these below:

Change in Species

As a practical policy, we suggest (Saxena, 1989a) that social forestry projects should be extended to reserved and protected forest lands by changing the nature of species from teak, eucalyptus and pines to usufruct-based trees such as *neem*, *mahua*, *tendu*, *sal*, *arjun*, *palmyrah* and tamarind. These should be supplemented with shrubs and bushes to yield fuelwood and fodder in the shortest possible time. This would strengthen women's access to forests and therefore benefits would be directly appropriated by them. Unlike commercial timber species, relatively low-value non-rotational trees for recurrent products would not attract the attention of powerful *panchayat* leaders and contractors.

We have analysed elsewhere (Chambers et al., 1989) why the social forestry programmes failed to achieve the social objectives of meeting local needs. Several other studies make the same point (CSE, 1985; USAID, 1985; Molnar, 1986; Arnold et al., 1987 and 1988; CIDA, 1988; ILO/ARTEP, 1988). One of the main reasons for this failure was that the implementation of social forestry programmes in India was heavily influenced by the perspectives of foresters and of foreign experts who advised the Government of India and the donor agencies. Both categories, because of their training and experience, had looked upon trees as timber, with profits to be obtained after felling. Therefore, even in the social forestry programmes, market-oriented species were planted.

The traditional Indian way of looking at trees has, however, been different. As opposed to trees for timber, Indian villagers for centuries have depended on trees for their livelihood. There has been little felling. Instead, trees have been valued for the provision of intermediate products which sustain and secure the livelihoods of the people.

The difference can be understood by comparing how fuelwood species are viewed in the two perspectives. According to foresters, fuelwood is obtained by felling trees having a high calorific value or as a by-product from lops and tops of timber trees. Casuarina and eucalyptus therefore seem perfectly justified on public lands. But poor women obtain fuelwood from twigs and branches of living trees and not by felling trees and often get little from the felling of so-called fuelwood trees. Casuarina and eucalyptus may be justified on farm lands if they improve farm incomes on a sustainable basis. But these trees hardly serve the poor when they are raised on public lands.

Given the inefficiency of administration and 'soft' character of the political system, one could generalise that out of a tree (on public lands) the stem goes to the rich and the towns, whereas branches, leaves and twigs belong to the poor. Therefore, the strategy should be to opt for species which have high proportions of branches and twigs relative to stem wood.

Thus, there is a world of difference between the plantation of eucalyptus and that of *prosopis* on roadsides. Eucalyptus benefits urban markets and industry, whereas *prosopis* not only solves the fuelwood problems of poor families but would also generate self-employment for women who prune the branches and sell them in urban areas. What is significant about *prosopis*, *ber*, *neem*, *karanji*, *agave* or other trees of low-market value but high biomass is that they do not require decisions about market shares between the rich and the poor. Their usufruct is not of much interest to the rich or is available to them only through gathering by the poor, so that, by default, benefits are available to the poor.

This requires a complete reversal of the recommendations of the National Commission on Agriculture, 1976, which favoured commercial plantations on forest land and trees for consumption and subsistence on private land (Saxena, 1990). Our recommendation is that subsistence and consumption should be met from forest and common lands and market demand should by and large be met from private land. No doubt, farmers would also raise multiple purpose trees which meet subsistence needs as well as provide income. Using private lands for short-rotation products will permit the large area of forest lands to be used for long-gestation trees which enrich the environment and provide a range of products to the poor. For quick benefits to the poor, long-gestation trees would be supplemented with an understorey of bushes and shrubs so as to satisfy immediate needs of the poor. 'Scientific' forestry should therefore mean that wild fruits, nuts, MFPs, grasses, leaves and twigs become the main intended products from forest lands and timber a by-product from large trees like tamarind, jack fruit and *sal*. The reverse has been the policy for the last 100 years. The maintenance of life support systems is a function performed mainly by the crown biomass of trees. It is this component of trees that can contribute positively towards the maintenance of the hydrological and nutrient cycles. Therefore, trees which provide a lot of leaves, twigs, and branches enrich the soil much better than those which provide poles and timber alone. Conventional forestry based on clear-felling disrupts the annual circulation of nutrients and increases soil erosion. Monoculture plantation forestry is prone to pest attack. Thus, in Kerala eucalyptus plantations raised after clear-felling dense green and deciduous forests have been devastated by fungal diseases and the consequent low productivity defeated the very purpose for which they had been raised (Nair, 1985).

Women have suffered in social forestry programmes because they are not organised and bureaucracy has failed to deliver their share of profits from sale of timber and pulpwood trees to them. Usufruct-based trees overcome both these problems. Under the new policy suggested by us, technology produces an output which eliminates the need for political and bureaucratic fine-tuning.

Sharing of Management

A further advantage of planting 'female' trees (which are essentially employment-augmenting trees as they require labour for gathering and collection) as opposed to 'male' trees (which are clear-felled) on forest and revenue lands is the likelihood of improved cooperation. People are reluctant to protect trees which will be auctioned or clear-felled to the benefit of the government, the contractors and the forest staff. They are much more likely to collaborate in the protection of trees from which they, much more than others, are in a position to benefit.

Social conditions in India are such that neither cattle nor human beings can be totally stopped from entering forests or village lands. What is therefore needed is adopting policies which improve productivity of degraded lands, living with constraints of the human and livestock pressure on such land. Given the ease of access to forests it has been possible, in practical terms, for the Forest Department to enforce its property rights. Forest lands, like revenue lands, have been a victim of the 'tragedy of the commons' phenomenon where community rights and management have not existed or have broken down. Therefore, any effort towards reforestation can yield results only if it involves local people, specially women, in planting and protecting trees. Sharing of management or usufruct or both with women, if properly implemented, could have wide-ranging implications on forest regeneration and welfare of the poor. Women's subsistence compulsions have shown greater sensitivity to ecologically sustainable development needs. Thus, it would be in the interest of the Forest Department to give higher priority to women's needs.

At least four different models of people's participation have been tried out by government programmes in India. The first are called social security schemes in which the poor, as individuals, raise and protect trees on government land with full wages, but get a share in the final produce too. The other three models are where protection is done jointly by a village, as in Arabari (West Bengal), where the responsibility for protection is handed over to a village society on payment of royalty and the village society distributes grasses to individual members, as in several villages around Sukhomajri in Haryana and lastly, where forest lands are leased to the rural poor, as in the Bharatiya Agro-Industries Foundation (BAIF) project in Vansda *taluka*, Gujarat, in which total control over forest lands got alienated from the Forest Department in favour of the people and the NGO. The last three models are well-documented and are not being described here.

Social security models

Several states like Maharashtra, Gujarat, Andhra Pradesh,

Rajasthan, Madhya Pradesh and Karnataka have introduced social security schemes. These schemes have at best provided wage employment but have not succeeded in instilling a sense of identification with government forests in the minds of the poor.

In 1988 I visited a number of sites in Andhra Pradesh to find the reasons (CIDA, 1988). First, nowhere had any agreement been executed or any letter given to the poor establishing their claims to usufruct. We were told that the agreement deed was pending with the Law Department. However, informally, many officers feared that signing a written agreement may be against the spirit of the Forest Conservation Act. Second, there was no continuity in wage employment. Although the scheme started in 1984, very few workers were at the site from the beginning. Third, each year the workers were shifted to a new site. They did not work on a sequence of contiguous plots as was envisaged in the scheme. Fourth, discrepancies existed from district to district on what was told to the beneficiaries and the degree of effort made to accommodate local needs. Some District Forest Officers permitted cultivation of legumes as an inter-crop but some did not, although the State Government scheme specifically permitted this. Fifth, species were selected, such as eucalyptus and *acacia auriculliformis*, which did not produce any intermediate goods of substantial benefit to the poor. On the other hand, people were so poor that women were forced to use leaves of these trees for cooking though the fumes harmed their eyes. This also removed litter from the ground, thus increasing the possibility of soil erosion. Their sense of identification with the planted trees would have been stronger had more socially useful trees been planted. The result of all this was that the beneficiaries saw themselves only as wage-employees working for less than minimum wages. The part of the deal concerning a share in the final produce meant little to them.

Looking at the practical conditions we feel that transferring complete decision-making, control and management over forest lands on a large scale to the *panchayats* would become feasible in the long run if the capability of village organisation is improved first through joint management schemes, as in Sukhomajri and Arabari (both being government and not NGO inspired schemes). The number of committed NGOs with vast organisations, which could be trusted with the management of forest lands, is limited. What is easier to achieve, given the present stage of *panchayats*' capabilities, is to enable people to have usufruct of intermediate and final products rather than try to establish immediately communal ownership over forest lands.

Moreover, the best organisation of management will depend on local conditions; and the balance of advantage to the poor as between Forest Department management and *panchayat* management of trees will vary. It is by no means automatic that the poor will be better off with *panchayat* management, especially if it entails trees for commercial pulpwood, poles for urban fuelwood, as shown by the panchayat-run woodlots in Gujarat and Tamil Nadu. In backward and tribal south Bihar, 'concepts like representation, participation and decision-making were completely alien to the rural women of the underprivileged communities' (GOB, 1987: 13). What was crucial to them was better access to forests for collection of fuelwood and minor forest products. Transferring total management of forest lands to the *panchayats* or other forms of people's organisations may be an ideal goal but the process will proceed at different places in different conditions. In the meantime, what is essential is to develop mixed forests in place of industrial plantations on forest lands to meet the livelihood needs of the poor. It is more pragmatic to change technology than institutions or ownership patterns.

Primacy of Rights

As already pointed out, women's rights of access and their right to collection of forest products are restricted, vague or not known to them. Sharing arrangements in social security plantations or community forestry schemes are ill-defined or not publicised or poorly implemented. There appears to be general reluctance on the part of governments to define clearly what people are to get, at what time and at what price in exchange for the participation expected of them. But participation of the poor and women is improbable unless their benefits are secure.

Therefore, we suggest that outside each forest coupe or social forestry plantation there should be a notice board publicising what rights people have as regards collection. The colonial tradition of secrecy must be given up. A simple notice that 'these trees belong to the community, and not to the government' may, in itself, change people's attitude towards *panchayat* plantations. Where rights are individualistic in nature, as in social security schemes, agreements must be entered in writing with the beneficiaries, informing them about their entitlement and copies given to each beneficiary.

Similarly, nationalisation of MFPs, justified in the name of preventing tribal exploitation, creates excellent opportunities for private traders and bureaucracy to extract 'suitable' rents for 'services rendered'. Practical considerations point out that the government is incapable of effectively administering complete control. It is better for the government to regulate private trade and to act as a watchdog rather than try to eliminate it. Monopoly purchase by the government requires sustained political support and excellent bureaucratic machinery. It is difficult to ensure these over a long period and hence nationalisation has often increased the exploitation of the poor. We suggest that for marketing MFPs, the government should not have a monopoly. The solution is to set up promotional marketing boards as distinct from commercial corporations (which are inefficient and hence demand nationalisation), with responsibility for dissemination of information about markets and prices to the gatherers. The boards would help in bridging the gap between what consumers pay and what gatherers get. We would be quite happy if government organisations could compete in the open market, as in the wheat purchase scheme, but the government should never acquire a monopoly. Nationalisation reduces the number of buyers and does not help gatherers in the long run. Setting up processing units within the tribal areas is also to be recommended.

Other Suggestions

Besides promoting usufruct-based trees, sharing management and giving publicity to their rights, the following measures need to be taken:

—There should be strict enforcement of labour laws, especially as regards maximum and equal wages, both when women work as wage-employees on nursery and plantation sites and when they are employed by contractors to collect *tendu* leaves and other MFPs. The government should take a lead in this respect and be a model employer for its own departmental works.

—Given the sex-segregated and hierarchical nature of Indian society, separate women's organisations and staff are needed to work among women to instil confidence in them so that they will fight for their rights.

—Each social forestry project should specifically consider the role of women at the project design stage and state in what way the project is going to help them, apart from exploiting their cheap labour. A project should aim at giving women new skills like training them in operating nurseries.

—More women need to be recruited to executive positions in projects. Without this, even the existing orders may not get implemented.

—Discriminatory land laws which prohibit women from owning land should be changed. A beginning can be made by making women owners of land where new tenurial rights are being created, as on wastelands.

—The government should learn from the experience of such NGOs which have achieved excellent results in helping women to become self-reliant through forestry. But the basic responsibility of providing equality of rights and opportunity to women through wastelands must rest with the government and not with NGOs.

THE NEW FOREST POLICY

Two problems will be encountered if the suggestions contained in this paper are to be implemented. One is the attitude of the civil servants, who associate 'development' with spending of money or with setting up of new organisations (which they head, of course!). Changes in policy or nature of species or laws are not seen as integral part of the development process because these have no direct financial implication. Non-monetary inputs in policy unfortunately have no ready acceptability in the government. The

Indian civil servant still has to learn the difference between planning and budgeting. He is looking for a scheme rather than a new policy framework. The question, however, is whether we wish to help hundreds of women through projects or millions of women through changes in policy.

The second is the new Forest Policy and the amended Forest Conservation Act, both of which were enacted in December 1988. There are two provisions of the amended Act which need to be taken note of. First, it bans assignment or lease of forest land to the people; and second, it prohibits plantation of horticulture crops, palms, oil-bearing and medicinal plants on forest lands unless prior permission of the Government of India has been taken. These provisions should be read with the new Forest Policy which looks down upon gathering of forest produce by stating that 'these (MFPs) and substitute materials should be made available through conveniently located depots at reasonable prices'. It also wishes to curtail the rights and concessions of forest-dwellers by relating them to the carrying capacity of forests.

The operative Section 2 (iii) of the Act (after amendment in 1988) reads as follows:

> No State Government or other authority shall make, except with the prior approval of the Central Government, any order directing that any forest land or any portion thereof may be assigned by way of lease or otherwise to any private person or to any other authority, corporation, agency or any other organisation not owned, managed or controlled by government.

This new section contradicts Section 28 of the Indian Forest Act, which states:

> Formation of Village Forests—(1) The State Government may assign to any village community the rights of Government to or over any land which has been constituted a reserve forest or declared a protected forest or is a forest belonging to the Government and may cancel such assignment. All forests so assigned shall be called village forests.

The new amendment nullifies the entire concept of village forests provided for in the Indian Forest Act. The rationale for keeping Section 28 in the Forest Act was the realisation that villages can better protect lands which are close to the village, and only land far away from habitation was to be brought under government control. In pursuance of the powers given to the States under this section, some States had transferred forest areas to village bodies for protection and management. These rules too have now become null and void and thus the existing legal basis for encouraging people's participation has been wiped out.

A similar ambiguous position exists regarding the choice of species. The new Act discourages planting of horticulture crops, oil-bearing plants, palms and medicinal herbs on forest lands. Although there is a good case for banning crops like oil-palms in the Andaman and Nicobar Islands or apples in Himachal Pradesh on forest lands yet, reading this clause with Section 2 (iii) gives rise to a suspicion that all usufruct-based trees, which attract people to the forest areas, like *ber*, *mahua*, *neem*, *karanj*, *palmyrah*, tamarind, *jamun* and several medicinal herbs would not be permitted now. This shows total insensitivity to the needs of the tribals and forest-dwellers. Already a large number of such livelihood trees have been replaced by monocultures of teak, eucalyptus and pines. This trend would get further support from the new Act.

Although it appears that these new draconian laws are not being implemented rigidly, as usufruct sharing schemes are being promoted by the Government of India itself, and the Government of India has also clarified that fruit species, etc., could be permitted, that power of harassment of NGOs by state officials under the garb of enforcing the Forest Conservation Act is immense. Further, who knows whether a more 'conservation-oriented' bureaucracy at the Centre tomorrow would not wake up to realise the powers given to them by the new Act! Further, as is well known, the legal enactments reach all sections of bureaucracy but benign circulars favouring people have a habit of disappearing fast before they reach everyone. This will further delay the execution of people-oriented schemes in the forestry sector.

Para 4.3 in the new Forest Policy relating to making forest produce available to the people from government depots amounts to banning the entry of the people to forest lands. It can be criticised on three grounds. First, it takes away the fundamental and age-old rights of the people to gather. India's forests are not uninhabited wildernesses; there are people living inside forests who, under the new dispensation, would have to travel miles outside their habitat to the government depot to buy essentials, which so far they got free. Second, the new scheme further strengthens the petty officials, who would get new powers to harass the forest-dwellers. And third, given the poor condition of State funds, the scheme would not be able to get enough subsidy and hence supplies are unlikely to be available at a price the poor can afford.

The objective seems to be that the government wishes to have exclusive monopoly rights to management and ownership over forest lands, both reserve and non-reserve. It does not trust the poor and hence would not permit growing of trees like *mahua* and mango which attract the poor to forest areas. The mid-term evaluation of Social Forestry Projects by a World Bank/USAID team revealed that even where the programme for regeneration of wastelands through tree *patta* schemes is unintended on road strips (as in Uttar Pradesh where road strips have been declared as protected forests) or unclassed forests in Himachal Pradesh, the programme has been held back due to lack of permission from the Government of India (USAID, 1988).

The new Forest Policy expresses concern for ecology but read with other provisions of the Act it appears that the environment is being used as a new excuse to keep people out of forests just as 'vital industrial needs' were considered enough of a justification to deny the legitimate aspirations of the poor in the past.

It is unfortunate that, while other departments are trying to give new assets to the poor, the Forest Conservation Act wishes to reduce access of the poor to whatever they could get from forests in the past. By denying to the people any role in forest management and a share in forest produce, it reduces the concept of people's participation to mere rhetoric. These tend to define forest-people interaction as a zero-sum game; forests can be protected only when people lose and any gain to the people is at the cost of forest protection. In this scenario, both lose. The challenge is how to convert this into a win-win game (Saxena, 1988). This is possible only when, as Shiva (1988: 64) argues, 'productivity', 'yield', and 'economic value' are redefined in terms of multipurpose utilisation and satisfying basic human needs. This requires a new outlook and a new strategy in which women's interests of secure rights of gathering (which are just as well the interests of all poor people) would be paramount. The livelihood needs of the poor are preconditions for sustainability of natural resources.

REFERENCES

Agarwal, Anil (1987) 'Between need and greed—the wasting of India: The greening of India.' In Anil Agarwal, Darryl D'Monte and Ujwala Samarth (Eds.) *The fight for survial* . New Delhi: Centre for Science and Environment.

Agarwal, Bina (1988). 'Who sows? Who reap? Women and land rights in India.' *Journal of Peasant Studies*, Vol. 15, No. 4, July.

———(1989) 'Rural women, poverty and natural resources: Sustenance, sustainability and struggle for change.' *Economic and Political Weekly*, 28 October.

Arnold, J.E.M. et al. (1987) Evaluation of the SIDA supported social forestry programme in Orissa, New Delhi.

———(1988) *Forestry for the Poor?* Stockholm: Swedish International Development Authority.

Brokensha, David (1988) 'Village-level management of common property resources, especially fuelwood and fodder resources in Karnataka, India.' A Report prepared for the World Bank, (mimeo).

Cecelski, Elizabeth (1985) *The rural energy crisis, women's work and basic needs: Perspectives and approaches to action*, Geneva: International Labour Organisation.

Chambers, Robert, N.C. Saxena and Tushaar Shah (1989) *To the hands of the poor: Water and trees.* New Delhi: Oxford and IBH.

CIDA (1988) 'Andhra Pradesh Social Forestry Project: Mid-term review.' New Delhi: Canadian International Development Agency.

CSE (1985) *The state of India's environment 1984–85: The second citizen's report.* New Delhi: Centre for Science and Environment.

Dasgupta, Subhachari and Ashoka Kumar Maiti (1986) *The rural energy crisis, poverty and women's roles in five Indian villages.* Geneva: International Labour Organisation.

Ellis, Frank (1988) *Peasant economics: Farm households and agrarian development.* Cambridge, UK: Cambridge University Press.

FAO (1989) *Households food security and forestry: An analysis of socio-economic issues*. Rome: FAO.

Fortmann, L. (1983) 'A role for women in agroforestry projects.' *Training for Agriculture and Rural Development,* Rome: FAO Economic and Social Development Series.

Fortmann, L. and D. Ricge Keay (1985), 'Women and agroforestry: Four myths and three case studies.' *Agroforestry Systems*, Vol. 2, No. 4.

FSI (1988) *State of India's Forestry, 1988* . Dehradun: Forest Survey of India.

GOB (1987) *Study on the participation of women in the Social Forestry Project*. Patna: Forest Department, Government of Bihar (mimeo).

———(1990) *Outlines of the new Social Forestry Project*, 1991–1996. Patna: Forest Department, Government of Bihar.

GOI (1976) *Report of the National Commission on Agriculture*, Part IX, Forestry. New Delhi: Ministry of Agriculture. Department of Agriculture and Cooperation, Government of India.

———(1987) *Background Note prepared for the Seminar on Welfare and Development of Tribals*. Ministry of Tribals, Government of India (mimeo).

GOI (1988) *Sharamshakti: National Commission on Self-Employed Women and Women in the Informal Sector*. Department of Women and Child Development. Government of India.

———(1988a) *Commodity Studies* (in six Volumes). Progressive Agro-industrial Consultants. Ministry of Welfare, Government of India.

GOMP (1987) *Forestry Project of Rehabilitation of Degraded Forests and Afforestation in MP*. Madhya Pradesh Forest Department, July.

GOO (1987) *Evaluation of Orissa Social Forestry Project*, Phase I. Bhubaneswar: Government of Orissa.

———(1988) *Social Forestry Project Document, Phase II* (1988–93). Bhubaneswar: Department of Forests, Government of Orissa.

Gupta, Tirath and Amar Guleria (1982) *Some economic and management aspects of a non-wood forest product in India: Tendu leaves*. New Delhi: Oxford and IBH.

IIM (1985) *Impact of Social Forestry Project on locals: A case study in Baduan district, UP*. Ahmedabad: Centre for Management in Agriculture, Indian Institute of Management (mimeo).

IIPO (1988) *Survival Rate of Trees: 1983–84 to 1987–88*. New Delhi: Indian Institute of Opinion Poll.

ILO/ARTEP (1988) *Employment and income generation through social forestry in India. Review of issues and evidence*. Report prepared for the Government of India and the Swedish International Development Authority (SIDA), New Delhi.

Jain, Shobita (1988) *Case studies of farm forestry and wasteland development in Gujarat, India*. Rome: FAO

Kulkarni, Sharad (1989) 'Forests: Law versus policy.' *Economic and Political Weekly*, 22 April.

Malmer, Pernilla (1987) *Socio-economic change in social forestry: A case study of Kovilur village, Tamil Nadu, India*. Uppsala: The Swedish University of Agriculture Sciences. Department of Economics and Statistics.

Molnar, Augusta and Gotz Schreiber (1989) *Women and Forestry: Operational Issues*. Washington. D.C.: Publications and Human Resources Department, World Bank.

Molnar, Augusta (1986) *Social forestry experience in India and Nepal: A review of community woodlots, the involvement of women and the introduction of wood-saving devices in the various projects*. Washington. D.C.: World Bank, ASPAB.

Nair, C.T.S. (1985) 'Crisis in forest resource management'. In J. Bandyopadhyay et al. (eds.), ***India's environment: Crises and responses.*** Dehradun: Natraj Publishers.

NWDB (1985) *Wasteland development.* National Wasteland Development Board, Government of India (mimeo).

Olsson, Gumilla (1988) *Employment and income generation in social forestry: A case study from Orissa.* Geneva: International Labour Office.

Pant, M.M. (1980) *Employment of women in forestry*, presented at the Seminar on Role of Women in Community Forestry. Dehradun: Ministry of Agriculture. Department of Agriculture and Cooperation, Government of India (mimeo).

PEO (1988) *Evaluation of the Social Forestry Projects in India.* New Delhi: Programme Evaluation Organisation, Planning Commission.

Rocheleau, D.E. (1987) 'The user perspective and the agroforestry research and action agenda.' In H.L. Gholz (Ed.) *Agroforestry: Realities and potentials.* The Netherlands: Martinus Nijhoff Publishers.

Saxena, N.C. (1987a) 'Women in forestry.' *Social Action*, Vol. 37, April–June, New Delhi.

———(1987b) *Commons, trees and the poor in the Uttar Pradesh hills.* London: ODI Social Forestry Network Paper 5f.

———(1988) 'Wrong Move.' *Illustrated Weekly of India* . 11 December.

———(1989a) *Degraded lands in India: Problems and prospects* . Bangkok: FAO.

———(1989b) 'Forestry and rural development.' *The Indian Journal of Social Science*, Vol. 2, No. 2.

———(1989c) 'Wasteland development: Development, environment protection and the poor.' *Indian Journal of Public Administration.* July–September 1989.

———(1990) *Farm forestry in North-west India.* New Delhi: Ford Foundation.

Shiva, Vandana (1988) *Staying alive: Women, ecology and development.* London: Zed Books.

Tripathi, S.L. (1988) 'Wastelands management in UP—Problems and prospects.' Paper presented at the BAIF Workshop on Wasteland Development, Pune, April 1988 (mimeo).

USAID (1985) *Second mid-term evaluation report: Madhya Pradesh.* New Delhi: USAID.

——— (1988) *National social forestry mid-term review.* New Delhi: USAID.

8 *Conclusions of the Workshop*

PREAMBLE

Peasant women have an intimate relationship with natural resources. This is based in part on culturally and socially defined gender-based roles which define their primary responsibility for collection of fuelwood, fodder, food and medicinal produce from the forests and other common property resources. This intimate relationship is inevitably and dramatically altered by changes in the environment and ecology as well as legal and policy-based changes in rights to access or collection, demographic changes and changes in the constitution, membership and responsibilities of local institutions. Hence the need to take into consideration the impact on women, for example, of both deforestation and afforestation programmes, and of the implementation of decentralisation policies. Strengthening the role of women in wasteland development offers an unusual opportunity to strengthen women's traditional relationship with the natural resource base and at the same time a means to enhance their status and address their strategic gender needs. For example, by ensuring their benefit and gain in programmes aimed at redistribution of land or other natural resources, and their equal representation in local bodies or development of alternative local women's institutions, peasant women's role can be strengthened. Wasteland development and management of local forest resources also offer significant potential for employment and income generation, especially for women because of their traditional roles in peasant and forest livelihood systems. Protection and regeneration of the natural resource base results in increased productivity as well as better fulfilment of subsistence needs and employment and income, reducing pressures on local populations to migrate in search of livelihood. Wasteland development should therefore be viewed as part of an integrated human and natural resource development strategy with the objective of creating a better overall quality of life for women, men and children, ensuring each of them a rightful place in society. Within this broad conceptual framework, the workshop participants arrived at the following conclusions:

Women and Local Institutions

1. In order to acknowledge and ensure peasant women's close association, dependence and concern for just and proper management, access to and control over forest resources, it is essential that such women constitute the majority of members in all legally constituted bodies for management of forest resources (e.g., Forest Protection Committees, Standing Committees of Panchayats responsible for forests reform and management, Van Panchayats, Village Forests Committees, etc).[1]

2. In order to ensure the effectiveness of new and forthcoming measures to expand women's participation in democratically constituted statutory bodies at the grass-roots (e.g., *panchayats*), it is imperative to initiate a process of organising and training women for empowerment, and to work towards gradual weakening of social resistance and constraints to their full and equal participation. To this end, the workshop recommends the following measures:

(*a*) Women's groups constituted by different schemes or departments of government (e.g., DWCRA, Women's Credit Corporations) or by NGOs should have a clear legal status. To facilitate this objective, simpler mechanisms and frameworks for legal registration are required than those presently available.

[1] *Editors' note*: One participant commented after the conclusions of the workshop were drafted that it was essential that women constitute the majority of members in all legally constituted bodies for management of forest resources but 'forest' areas are not the only issue. The term 'wasteland' incorporates even village common lands, revenue and private wastelands.

This is a very important point which was not discussed at the workshop but needs to be mentioned here. Forest lands have institutional structures with representative bodies of various kinds where women's participation can be ensured. But similar legal institutional structural possibilities are not available for 'wastelands'.

(*b*) Since the Government of India has already ratified ILO Convention 141 concerning Rural Workers Organisations, the Government of India and the ILO should jointly appoint an Expert Group to examine existing laws for registering organisations, of the rural poor (e.g., Cooperatives, Registered Societies, Trade Unions, etc.) to see whether they need modifications in the changed context of today's development, including the multiple roles such groups are playing (or are required to play), to suggest needed amendment or modification of such legislation to ensure effective participation of women in such bodies both as members and as managers.

(*c*) The Department of Rural Development, Government of India, should consider implementing the recommendation of the Committee on the Status of Women in India (1974) on establishing women's *panchayats*. This could be considered as a support measure to the present Constitutional Amendment Bill regarding *panchayats* that was before Parliament.

3. Recommendations of the committee constituted by the Home Ministry on the Code of Grants to Voluntary Agencies should be disseminated for public review and discussion with a view to facilitating the flow of funds to rural women's organisations involved in wasteland development.
4. Such bodies which require representation of all participating households in the village should have one male and one female from each household. Other nominated executive bodies related to management of natural resources should consist of a majority of women. Elected bodies should endeavour to ensure equal representation of men and women.
5. In forest areas, where women are the primary collectors of fuel and fodder, forest protection and management committees should have a membership composed primarily of women.

Employment and Income

6. Under wasteland development programmes and in work carried out under forest departments or corporations, men and women should be paid at least the minimum wage, and women should be paid equal wages to men for work of a similar nature. These minimum standards must be strictly enforced as per

International Labour Convention 26 (1928) concerning Minimum Wage-Fixing Machinery, Convention 100 (1951) concerning Equal Remuneration and Convention 111 (1958) concerning Discrimination (Employment and Occupation) all of which have been ratified by India.

7. Projects aiming at the protection and regeneration of the natural resource base should link up to existing government schemes such as TRYSEM, DWCRA, etc. These schemes, which are designed to promote employment and income generation in rural areas, should also take up these activities in a systematic way, ensuring linkages to existing technical institutions and programmes such as services available from the agricultural universities.

Technology and Training

8. Alongside wasteland development for meeting fuel needs of people, emphasis on development of viable alternate sources of energy available and accessible to the rural poor should be given priority. This should be promoted hand in hand with employment and income generation efforts.

9. The whole infrastructure for training in agriculture and forestry-related activities of both the government (e.g., agricultural universities, Krishi Vigyan Kendras, TRYSEM, etc.) and NGOs should be used more equitably and effectively. With a view to developing a more effective methodology for gender sensitisation of development research, training and policy personnel on the one hand, and needed knowledge and technology transfer to women on the other, more dialogue between women at the grass-roots level and development researchers, trainers and policy-makers should be actively promoted.

10. It is recommended that technical training and assistance should be given according to different ecological zones across the country for both voluntary organisations and government staff who are working at all levels in wasteland development. This might be done through whatever institutions have the concerned expertise and could be coordinated by such national institutions as the NWDB or the NIRD.

11. There should be special provisions for wasteland development in arid zones and areas with special problems which require more funds and a different budgeting structure due to the harsh conditions prevalent as compared to other areas.

12. More extensive technical support such as soil testing, engineering and agricultural extension is required by NGOs and village-level *mandals* involved in wasteland development work. Professionals involved in this work also require training in technical aspects.

Information Dissemination

13. Rural women are by and large illiterate and ignorant of their legal rights—rights to own land, lease and tenurial rights, usufruct rights to non-timber forest product, etc. Existing documentation on legal rights related to land and local institutions such as those prepared by PRADAN (New Delhi), the National Centre for Human Settlements and Environment (Bhopal), MARG (Delhi) and the ILO should be widely disseminated. A publication grant, including provision for translation into local languages, may be recommended to the Planning Commission.

14. In order to build up an effective information network among women engaged in wasteland development, exchange visits, workshops and study tours for NGOs and village women should be supported to enable local groups to exchange experience and learn from each other.

15. It is recommended that the government establish promotional marketing boards for minor forest produce and other forest products (see 16 below). These boards would assume responsibility for dissemination of information on markets and prices to the gatherers.

Policy Issues

16. The government should not acquire monopoly rights over buying forest products from the gatherers as nationalisation keeps prices low by reducing the number of buyers and does not help the gatherers in the long run. The government should instead provide a minimum guaranteed price for each forest product, and should compete with private trade. This could be done through promotional marketing boards, as suggested above (15), which inform the public about markets, and prices so that the gatherers could get fair market prices.

17. Social forestry should be extended to reserved and protected forests. The nature of species promoted should be changed from commercial, largely monoculture crops like teak, eucalyptus and pines, to usufruct-based trees such as *neem, mahua, sal, arjun,* tamarind and others which are local species. 'Scientific' forestry should therefore mean that water, wild fruits, nuts, MFPs, grasses, leaves and twigs become the main intended products from forest lands and timber a by-product. Commercial species may be promoted through farm forestry on private lands.

18. The National Wasteland Development Board (NWDB) should be revitalised, and should resume dialogue with NGOs, academics and other organisations interested in wasteland development. The recommendations of these experienced organisations should be taken into account while formulating policies and widely disseminated for public discussions. The NWDB should be allowed to develop as an autonomous organisation that can take an independent view, if necessary, of the government in power, and provide continuity of analysis, policy formulation and implementation strategies.

19. Land used by tribals for shifting cultivation, now classified as government property, should be brought under a system of joint management between the tribals and the government.

20. A committee consisting of government representatives, grassroots workers, academics and NGOs should be set up to recommend the mode of sharing of produce from forest and public lands between the people and the Forest and Revenue Departments.

Legal Rights and Legislation

21. The Indian Forest Act, the Forest Conservation Act and the Forest Policy should be suitably amended to:

(*a*) recognise gathering from forest and public lands as a fundamental right of poor women and men;
(*b*) promote people's participation by assigning to village communities and bodies of women forest lands which would then be regenerated with their active participation, management and control; and
(*c*) encourage planting of mixed species of fruit trees, oil-bearing, food-giving and medicinal plants.

22. Land laws of States like Gujarat and Maharashtra provide that only agriculturalists can own lands. It should be clarified that the term 'agriculturist' includes agricultural labour and poor peasants, both men and women, including those who have lost control over their lands.
23. The rights of local people to collect non-timber forest produce should be legally secure and protected.
24. The laws relating to land, water, agriculture and forests should be re-examined to assess their effects on women, to safeguard their interests and to ensure that there is no discrimination against women. Discriminatory clauses that affect women's equal rights to land and other natural resources in rural areas need to be removed in defence of women's constitutional rights to equality in law in all state-promoted measures, and in all state-created institutions and programmes.

Tenure and Allotment Priorities

25. Women's groups, including *mandals*, cooperatives, societies, etc., should be given priority in the allotment of revenue wastelands for development.
26. Longer leases, i.e., for at least 50 years, along with provision for renewal, should be given to women's groups by revenue departments, taking into consideration setbacks often caused by natural calamities and other factors in developing plantations and degraded lands.
27. Where new land allotments are being made to individual households (e.g., under land reform or housing programmes), ownership should be directly in the name of women.

Glossary

aadam	- Mitragyna parviflora
aalam	- A tree which is worshipped
adivasis	- Tribals
agave	- Agave sisalana
agyaras	- A village meeting traditionally attended by men in parts of Rajasthan
amla/amlaki/aonla	- Phyllanthus emblica
anantmool	- Medicinal herbs
angami nagas	- A tribal group
anganwadi	- Creches in courtyards
anjun	- Hardwickia binata
anwalia	- Emblica officinala
aonla/amla/amlaki	- Phyllanthus emblica
aranya samskriti	- The culture of the forest
aranyani	- Mother goddess
arhar/tuar	- Cajanus cajan or pigeon pea
arjun	- Terminalia arjuna
asan	- Terminalia termentosa
babul/baval	- Acacia nilotica
bacher	- Bark of a tree used for brewing liquor
bajri	- Pennisetum typhodeum or pearl millet
bamboo	- Dendrocalamus strictus
banjara grass	- A fodder grass
ber	- Zizyphus nummularia
bhabbar	- Eulaliopsis binata
bhanj	- Quercus spp. or oak
bhendi	- Hibiscus esculentus or lady's fingers
bhimal	- Grevia optiva
beedi	- Indigenous cigarette
bigha	- A unit of measurement of approximately 1000 square yards
bogodu	- Shifting cultivation
chaknni	- Medicinal plant
channa	- Cicer arietinum or bengal gram
charpoy	- Cot with jute ropes
chaula	- A local variety of bean
chernot	- Grazing lands or revenue wastelands in Rajasthan
chikku	- Achras zapota L.
chowkidar	- Watchman
chulha	- Cooking-stove
chulha mistris	- Cooking-stove mason
coolies	- Wage labourers

dais	- Traditional midwives
dakin	- Witch
datun	- Reeds used for cleaning teeth
desi baval	- Acacia nilotica
dhabdra	- Boswellia serrata
dhaman	- Dicanthium annulatum
dhawra	- Anogeissus latifolia
dhuna	- A gum from the tree shorea robusta
dungar	- Small hill
dungar dais	- Para-veterinarians
durree	- Carpet
eucalyptus	- Eucalyptus vr. astringens
falah/phala	- Hamlet
gameti	- Traditional village leader
ganda baval	- Prosopis juliflora
gauchar	- Grazing land
gomal	- Revenue grazing land
gotras	- Clans
gram	- Village
gram panchayat	- Elected village council
gram pradhan/pradhan/ sarpanch	- Elected head of the village council
guntha	- A unit for measurement of land equal to one-fortieth of an acre
harataki	- Terminalia sheubula
hukra	- A poor quality fodder grass
israeli babul	- Acacia tortillis
jack-fruit	- Artocarpus integrifolia
jamun	- Syzygium cumini
janggi	- Medicinal herbs
kalmegh	- Medicinal herb
karanj	- Pongamia pinnata
kendu/tendu/timru	- Diospyros melanoxycon
khadi	- Cloth made of handspun yarn
khair	- Acacia catechu
khatta imli	- Tamarindus indica
khedwa	- A resin used in the preparation of Indian sweetmeats
kisan	- Farmer
laha	- Lac, a gum from Butea monosperma
lodh	- Bark of a tree used for medicinal purposes
mahila mandal	- Women's group
mahila mangal dals	- Women's groups in the hills of Uttar Pradesh
mahila van panchayat	- All-women village-level bodies looking after the management of the forests

mahua	- Madhuca latifolia
mahua flowers	- Madhuca indica
mahuda	- Bassia latifolia
mamlatdar/tehsildar	- A sub-district level government official concerned with land records and revenue administration
mandali	- Group
mehendi	- Lawsonia inermis
mela	- Fair
mistri	- Mason
muttoga/palas	- Butea monosperma
nalas	- Gullies
namaal	- Seasonal migration
neem	- Azadirachta indica
nishtar	- Customary
oak	- Quercus spp.
padyatras	- Walking tours
palas/muttoga	- Butea monosperma
palmyrah	- Palmyrah indica
panchas	- Members of elected village council
panchayats	- Local elective bodies in charge of village development
panchayat samiti	- A developmental committee for a group of panchayats.
parma	- Medicinal herb
pars	- A tribal group
parura	- Bark of a tree used for medicinal purposes
patal gudu	- Medicinal herb
patta	- A document conferring title to land
patwari/talathi	- Village-level government official concerned with land records and revenue administration
pendula	- Anogeisus pendule
phala/falah	- Hamlet
pilu/piludi	- Salvadora piscica
pine	- Pinus roxburghii
pradhan/gram pradhan/ sarpanch	- Elected head of the village council
prant adhikari	- Sub-divisional Officer, a government official in charge of three or four *talukas*
puja	- Ceremonial worship
purdah	- The tradition of seclusion of women
ragi	- Eleusina coracana
rahara	- Terminalia balerica
rakhi	- A traditional Indian ceremony where the sister ties a string on the wrist of the brother. Symbolically, the sister is placed under the brother's protection
ram baval	- Tree from the acacia family

ram datan	- Medicinal herb
sabhapati	- Head of a developmental committee for a group of *panchayats*
sal	- Shorea robusta
samadhan	- Settlement
samiti	- Women's group
sanghas	- Groups
sarna	- Sacred grove
sarpanch/gram pradhan/ pradhan	- Elected head of *panchayat*
sarvodaya	- A social movement inspired by Gandhian ideals
satmool	- Medicinal herb of the asparagus family
semal	- Salmalia malabaricum
seras	- Albizzia lebbek
sewan	- Gmelina arborea
shibir	- Camp
shramdaan	- Voluntary labour
sona	- Acacia auriculiformis
soyam	- Community-owned pastures in the hills of Uttar Pradesh
subabul	- Leucaena leucocephala, a fast-growing nitrogen-fixing tree; yields both fodder leaves and fuelwood.
talathi/patwari	- A village-level revenue official
taluka/tehsil	- A geographical unit of the district
tamarind	- Tamarindus indica
tassar	- Silk tassar is a product of insects which are cultivated on the leaves of arjun and sal trees
teak	- Tectona grandis
tehsil/taluka	- A geographical unit of the district
tehsildar/mamlatdar	- A sub-district level government offficial concerned with land records and revenue administration
tendu/kendu/timru	- Diospyros melanoxycon
timru/tendu/kendu	- Diospyros melanoxycon
tuar/arhar	- Cajanus cajan or pigeon pea
urad	- Phoseolus mungo or black gram
van pals	- Protectors of the forest
van panchayat	- Village-level bodies looking after the management of the forest
vanputras	- Sons of the forest

About the Contributors

Neera Burra is National Expert for ILO's Women and Wasteland Development projects in Gujarat and West Bengal since 1989 in New Delhi, India. She was also Programme Officer for the Labour and Population Team for Asia and the Pacific for one year. After her Ph.D. in Sociology from the Delhi School of Economics in 1985, she researched aspects of child labour in many industries in India as a consultant to various organisations like UNICEF, UNESCO, DANIDA, ILO and ILO/ARTEP. Dr. Burra has published her findings extensively both nationally and internationally and her book on child labour is to be published.

Martha Alter Chen, a research associate at the Harvard Institute of International Development (HIID), is an experienced development practitioner, planner, and researcher, who has specialised in gender and poverty issues. Her overseas experience includes eight years in Bangladesh, where she worked for the Bangladesh Rural Advancement Committee (BRAC) as their executive assistant in charge of women's programmes, and seven years in India where, in her capacity as Oxfam America's field representative for India and Bangladesh, she worked closely with over 50 non-governmental projects. In both capacities, Dr. Chen was involved in formulating rural development policies, designing and managing rural development projects, and training NGO project staff and beneficiaries.

Dr. Chen has written extensively on development issues, including a book on BRAC's women's programme and the economy of a Bangladesh Village; household coping strategies in an Indian village; case studies of selected development projects for women; and policy documents on integrating women into mainstream development programmes. She is currently engaged in a study of widows in rural India with support from the World Institute for Development Economic Research

(WIDER). She co-teaches a course on poverty and policy in developing countries at the Kennedy School of Government, Harvard University.

Vasudha Dhagamwar is a teacher turned legal activist. She has done participatory research on land laws and criminal justice in tribal areas culminating in Supreme Court cases. She has been interested on issues concerning unorganised labour, displacement and rehabilitation of project displaced persons and women's rights. Dr. Dhagamwar is the Executive Director of the Multiple Action Research Group (MARG) in New Delhi.

Renu Khanna worked for eight years in the area of health care management through VHAI (Voluntary Health Association of India). The aim was to train administrators of voluntary health organisations to apply management principles to improve the functioning of their organisations. During the course of this, she cultivated skills as a trainer and an organisation development consultant.

In 1984, a voluntary agency called SAHAJ—a society for health alternatives—was registered as a group. As a member of this group she has been working as a resource person for grass-roots organisations for capacity building and on issues related to women and health, women and wasteland development, legal education for women, with empowerment as the ultimate objective.

As a free-lance management consultant, she undertakes short-term assignments like feasibility studies, reviews and evaluations.

Geeta Menon, a post-graduate in Sociology from the University of Delhi, has been working since 1983 on issues concerning tribals and forests in India. She has completed three major research projects on 'Forests, Environment and Forest-Dweller Economy in Orissa', 'Tribal Women and the Informal Forest-Dweller Economy' and 'Impact of Migration on Tribal Women's Work and Status in Orissa'. The first two were with the Indian Social Institute, New Delhi, and the third with the Indo-Dutch Programme for Alternatives in Development, New Delhi/Amsterdam (IDPAD). In addition, she has worked as consultant to USAID and has conducted several evaluations of voluntary organisations in Orissa and Tamil Nadu. She has co-authored three major books and written several articles for reputed journals and books published in India and abroad. She is presently working on a research project on Social Forestry examining efforts at the grass-roots-level with IDPAD.

Madhu Sarin graduated as an architect in 1967. This was followed by an assignment in urban development planning issues in the Third World, focusing on the problems of slums and squatter settlements and their relationship to urban land policies in India.

From 1980, she started working on rural natural resource management issues and women's relationship to them. While working on an innovative project in the villages of Sukhomajri and Mada in Haryana from 1980 to 1983, she developed an improved cooking stove in collaboration with village women. From 1983 to 1987, along with a small team, she developed a dissemination methodology for these stoves based on training local village women as skilled stove builders. The close contact with rural women in several north Indian states had her deeply involved in gender relations in natural resource management.

N.C. Saxena, a policy analyst, worked with the Government of India as Joint Secretary, Ministry of Environment and Forests. He is currently at the Oxford Forestry Institute, U.K., completing his Ph.D. on 'Trees on farm lands in north-west India'. His publications include a book, *To the hands of the poor: water and trees* written jointly with Robert Chambers and Tushaar Shah.

Chandrika Sharma has been working as a member of Sahaj, a small resource group, since 1986. As a member of this group she has worked with several rural based NGOs in Northern India on issues related to gender, environment and credit. She has been closely associated with the environmental and women's programme of PEDO.

Andrea M. Singh was Chief Technical Adviser of the ILO's Programme for Rural Women in South Asia. Based in New Delhi since 1971, she has carried out policy-oriented research on the urban poor, rural to urban migration, the informal sector including home-based workers, and the development of voluntary associations in India. She earned her Ph.D. in Anthropology from Duke University in 1972 and since then has authored numerous books and articles, including *Neighbourhood and Social Networks in Urban India* (1975), *The Urban Poor* (co-author, 1980), and *Invisible Hands* (co-editor, 1987). In mid-1991, she was transferred to Dar-es-Salaam, Tanzania, as Director of the ILO office.

Enakshi Ganguly Thukral has been working with MARG in New Delhi since 1986 and has been involved in studies on labour in the unorganised sector, displacement and rehabilitation of project displaced persons. Prior to joining MARG she was working with the Indian Social Institute.

Philip Viegas is a Post Graduate in Social Psychology from the Loyola University of Chicago. He has worked on two major research projects with the Indian Social Institute, New Delhi—'Forests, Environment and the Forest-Dweller Economy in Orissa' and 'Land Alienation among Tribals in Orissa'. Subsequently, he has been consultant with USAID on 'Private Sector Survey and Natural Resource Management in India' and has conducted evaluations of voluntary organisations in Orissa, Tamil Nadu and Maharashtra. He is the author of *Encroached and Enslaved: Alienation of Tribal Land and its Dynamics* and the co-author of two other books—*Forests, Environment and Tribal Economy* and *Impact of Environmental Degradation*. He has also co-authored a monograph on *Participatory and Conventional Research Methodologies*, and has several published articles to his credit. He is presently working for CARE-India as Evaluation Coordinator.

Index